Elmer W. Sundberg

Building Trades Blueprint Reading

5th Edition

ATP AMERICAN TECHNICAL PUBLISHERS, INC.
HOMEWOOD, ILLINOIS 60430

Part I Fundamentals

Copyright © 1945, 1949, 1956, 1967, 1972
by AMERICAN TECHNICAL PUBLISHERS, INC.

All Rights Reserved

Library of Congress Card Catalog No.: 67-16046
ISBN: 0-8269-0435-1

56789-72-1817

It should also be noted that in many cases the authors have indicated relative sizes, location and types of materials found in small commercial and residential construction in order to begin to give the student a feel for the scale of the elements which go into the making of such buildings. In actual practice, all sizes, types, and location of materials are subject to change based on structural analysis, local building codes, safety standards and local building practices.

WE, THEREFORE, DISCLAIM ANY RESPONSIBILITY FOR THE USE OF THE DRAWINGS AND/OR PICTURES PRESENTED IN THIS TEXT FOR ANY PURPOSE OTHER THAN THE TEACHING PURSUANT TO THIS PUBLICATION.

No portion of this publication may be reproduced
by *any* process such as photocopying, recording,
storage in a retrieval system or transmitted
by any means without permission of the publisher.

Printed in the United States of America

Note to the Instructor

Along with developments in the building trades, the techniques used in drafting are continuously being changed and improved year by year. Improvements are made so that the blueprint will provide a better means of communicating ideas between the architect and the building tradesmen. This book includes the new standards which are now in use as well as generally accepted practices for preparing working drawings.

Throughout the book Self-Check Quizzes are provided so students may test themselves on their progress. Answers are provided in the Appendix. Nine Trade Competency Tests and a comprehensive Final Exam are also included.

A complete set of blueprints for a contemporary home is given at the end of the book. The design characteristics, structural relationships, and the uses of materials are thoroughly discussed.

The 5th Edition has been revised to include:

- Updated Electrical Symbols (page 49). GFCI and SD added.
- Insulation symbol added to sections (pages 79 and 81).
- Floor trusses (page 84). Floor and roof trusses discussed under *Prefabricated Houses.*
- Latest Welding Symbols (page 144) AWS A2.4-79.

THE PUBLISHERS

Contents

Chapter 1	5	**Working Drawings and Blueprints**

Working Drawings—How Blueprints are Made—Self-Check Quiz No. 1.

Chapter 2	8	**Reading Drawings**

Pictorial Drawings—Perspective Drawings—Isometric Drawings—Oblique Drawings—Three View Drawings (Orthographic Projection).

	15	TRADE COMPETENCY TEST NO. 1
Chapter 3	17	**Reading Elevation Drawings**

Information to be Found on Elevation Drawings—Design of Building—The Roof—Self-Check Quiz No. 3A—Openings—Self-Check Quiz No. 3B—Dimensions—Exterior Finish: Materials—Exterior Finish: Trim—Miscellaneous Details—Self-Check Quiz No. 3C.

	29	TRADE COMPETENCY TEST NO. 2
Chapter 4	31	**Reading Floor Plans**

Importance of Floor Plans—Basic Ideas in Reading Floor Plans—Reading Floor Plans—One Story House—Two Story House—One-and-One-Half Story House—Self-Check Quiz No. 4.

	41	TRADE COMPETENCY TEST NO. 3
Chapter 5	43	**Symbols and Notations Used on Floor Plans**

Symbols for Material and Building Parts—Abbreviations—Self-Check Quiz No. 5A—Self-Check Quiz No. 5B—Self-Check Quiz No. 5C—Self-Check Quiz No. 5D—Self-Check Quiz No. 5E.

	61	TRADE COMPETENCY TEST NO. 4
Chapter 6	63	**Scaling and Dimensioning Practices**

Drawing to Scale—The Architect's Scale—Reading a Scale—Self-Check Quiz No. 6A—Using Folding Rule to Draw to Scale and to Read Dimensions—Measuring Blueprints to Obtain Dimensions—Self-Check Quiz No. 6B—Self-Check Quiz No. 6C—Drawing Symbols and Conventions to Scale—Dimensioning Standards—Exterior Walls—Interior Partitions—Dimensioning Methods—Windows and Doors.

	75	TRADE COMPETENCY TEST NO. 5
Chapter 7	77	**Reading Blueprints for Structural Information**

Finding Structural Information on Blueprints—Construction Types—Platform or Western Framing—Balloon Framing—Plank and Beam Framing—Brick Veneer Construction—Masonry Buildings—Building Using Unit Construction—Prefabricated Houses—Self-Check Quiz No. 7.

	87	TRADE COMPETENCY TEST NO. 6
Chapter 8	89	**Reading Detail Drawings**

The Scale Used for Details—Dimensions—Locating the Detail—Details of Interior Wall Elevations—Details of the Structure—Window and Door Details—Details of Exterior and Interior Trim—Self-Check Quiz No. 8.

	101	TRADE COMPETENCY TEST NO. 7
Chapter 9	103	**The Plot Plan**

The Survey Plat—Scales Used by the Surveyor—Elevations—Contour Lines—The Plot Plan—Self-Check Quiz No. 9.

	111	TRADE COMPETENCY TEST NO. 8
Chapter 10	113	**Reading Blueprints for Trade Information**

Types of Trade Information on Blueprints—Division of Labor—Specifications—Modular and Component Concepts—Electrical Work—Plumbing—Sheet Metal Work—Heat and Air Conditioning—Concrete Work and Masonry—Welding—Self Check Quiz No. 10.

	153	TRADE COMPETENCY TEST NO. 9
Chapter 11	157	**Reading a Set of Blueprints**
	165	**Final Exam**
	169	**Appendix A Reviewing Mathematics**
	175	**Appendix B Metric Conversions**
	177	**Glossary**
	181	**Answers to Self-Check Quizzes**

Chapter 1

Working Drawings and Blueprints

Everyone concerned with construction or with buildings should have some idea how the plans are created and how they are reproduced in the form of blueprints. The complex combination of lines, symbols, notations, and dimensions which appear on the blueprints should be understood so that the information can be used intelligently.

A *blueprint* is a reproduction of a drawing which has been made on very thin, translucent paper. Almost any number of copies or blueprints may be made using a process which resembles that used in making prints of photographs. A true blueprint has white lines on a blue background. However, some reproductions of drawings are made using a process which produces a black or blue line on a white background. All of these reproductions, regardless of the color of the lines or the color of the background, are called "blueprints."

When an architect prepares a set of drawings containing all of the information and dimensions necessary to carry a job through to successful completion, he has made a set of *working drawings*. Reproductions of these working drawings are called a set of blueprints.

Confusion results when a set of blueprints is called a set of working drawings or a set of architect's plans. The blueprints may be called by these terms correctly because they are exact copies. However it must always be understood that blueprints are reproductions. By following the steps in the development of a set of working drawings this matter can be made clear.

Generally the prospective owner will meet with an architect to discuss the planning of the house. The architect will want to know about the family so that he can provide for its needs. He will want to know about the lot; the materials, equipment, and fixtures desired; the approximate price range; and the preference in layout and design. He will then make a series of study sketches and present them to the client. His skill in planning and his knowledge of materials, of costs, and of building codes will be evident as he makes his presentation. When the prospective owner makes his final decision the architect is ready

Charles Bruning Company

Fig. 1-1

The diazo printing machine produces prints with black or blue colored lines on a white background.

to proceed with the working drawings. He uses a very good grade of tracing paper which will stand erasing and the abuse it gets as it is used over and over again in the blueprint making process. The tracing paper must also be very translucent so that as much light as possible will pass through it. Generally he works with pencil, skillfully drawing the lines so that they are uniform and opaque.

Working Drawings. The working drawings, or architect's plans, consist of several different kinds of drawings usually assembled into a set.

A. *Floor plans,* which are almost always drawn first, show the layout of rooms, information about windows, doors, cabinets, bath fixtures, and many other features of the building.

B. *Elevation drawings* show what the exterior of the house and the roof will look like. They will indicate whether the house is built of brick or stone or if the exterior walls are covered with siding, shingles, or plywood panels. Additional information about windows and doors will be shown.

C. A *section drawing* taken vertically through an exterior wall may be drawn on the same sheet as one of the elevation drawings. Details of construction and information about floor levels in relation to the grade are shown.

D. The *plot plan* tells where the house is to be located on the lot and gives other information about utilities, grading, walks, etc.

E. *Detail drawings* are usually drawn to a larger scale than the other drawings in order to show special features such as an entrance doorway, a fireplace mantel, or a built-in cabinet.

The set of working drawings is then sent out to the blueprinter who makes as many sets of blueprints as are needed by the contractors for estimating purposes and later for the actual construction.

How Blueprints are Made. Simply stated, blueprints are photographic copies of the architect's original working drawings. In 1840 a man named Herschel discovered a method of producing paper sensitized with iron salts which would undergo chemical change when exposed to light. The blueprint process had its start when drawings made on translucent paper were placed over the sensitized paper in a glass frame used to hold it firmly. The frame was then exposed to sunlight. A chemical action took place wherever the light was permitted to strike the paper. When the blueprint paper was washed in water the part protected by the pencil or ink lines on the tracing would show as white lines on a blue background. A fixing bath of potassium dichromate, a second rinse with water, and the drying of the print completed the operation. Blueprints are made today using the same steps but with high speed machinery. The method is relatively simple, and the prints are relatively inexpensive and quite permanent.

Many of the blueprints used today have blue or black lines on a white background. This is usually achieved in one of two ways, either by using a negative intermediate called a Van Dyke (or brownprint) or by using the diazo process.

1. A negative intermediate is made by exposing the original tracing with a sheet of

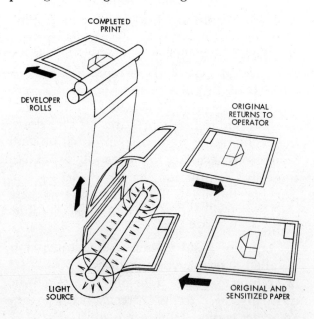

Charles Bruning Company

Fig. 1-2

The original tracing and sensitized paper are exposed to a strong light source. The sensitized paper goes on to be developed.

Van Dyke paper to strong light. After development the Van Dyke sheet becomes dark brown in color with translucent (white) lines. When a Van Dyke sheet is used to make blueprints instead of the original tracing, they will have blue lines on a white background.

2. Most of the blue line or black line blueprints are made today using the diazo process. See Fig. 1-1. There are two types of sensitized paper used; each one is developed differently. The papers are coated with a chemical which, when exposed to ultra-violet light, becomes a part of a dye complex. The original tracing is placed over a sheet of this sensitized paper and is fed into a copying machine. See Fig. 1-2. The two sheets revolve around a glass cylinder so that they are exposed to the light. The tracing is returned to the operator and the sensitized paper continues on to be developed.

When the moist diazo process of developing is used, the paper passes under a roller which moistens the top surface completing the chemical reaction to bring out the image. When the dry diazo process is used the sensitized paper is sent through a chamber in which its surface is exposed to ammonia vapor. The ammonia vapor precipitates the dye to bring out a blue line on the paper.

The diazo processes have the advantage of providing excellent reproductions with perfect accuracy because the paper has not been soaked with water and then dried. The blueprints resulting from the moist process have a black line on a white background. The blueprints resulting from the dry method have a black line on a white background. Thus white background prints are made without the extra expense of making a negative intermediate. White background prints are easier to read than conventional blueprints and notations or emergency corrections can be shown with greater clarity.

Self-Check Quiz No. 1

(Based on Chapter 1)

Fill in the blank spaces then check your answers. *Answers are given in the appendix.*

1. When the owner agrees to the architect's study sketches what is the next step? _____

2. What are blueprints? _____

3. When the architect draws the working drawings, what two qualities should the paper he works on have? _____

4. What basic types of drawings make up an architect's set of working drawings? _____

5. Conventional blueprints are made on paper sensitized with iron salts. After exposure to light how are they developed? _____

6. A conventional blueprint with blue lines on a white background requires a negative intermediate. What is it called? _____

7. Why is one diazo process called the moist process? _____

8. What is used to develop the image in the dry diazo process? _____

9. What are the colors of the lines obtained by using the two diazo processes? _____

10. Why are blue or black line prints preferred by some builders? _____

Chapter 2

Reading Drawings

Pictorial Drawings. It is easy for a person to understand what a house looks like when he sees it in picture form. There are several pictorial ways of showing a building which are helpful in learning how to read blueprints.

Perspective Drawings. The architect often makes a drawing called a perspective which has many of the features of a photograph, even though it is drawn long before the house is built. He shows the house from a particular vantage point which displays its most interesting aspects. He draws the shrubs, trees, walks, streets, and even indicates shadows as though the sun were shining on the house from a particular direction. Such picture-like drawings appear in the real estate sections of newspapers, in periodicals, and in brochures used to interest buyers. Whenever perspective drawings are available the student should use them to help him visualize the many details which appear on the working drawings. See Fig. 2-1.

Isometric Drawings. The isometric method

Fig. 2-1

A perspective sketch of a house is easily understood because it is like a picture. The architect chooses a vantage point which will show the most interesting features.

Reading Drawings 9

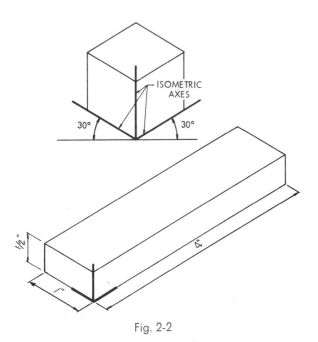

Fig. 2-2

Isometric Drawing: Three axes are laid out first; full measurements are used along each axis and any line parallel to one of the main axes.

offers a ready solution for the beginner who wants to create a three dimensional effect and for the experienced draftsman who wants to illustrate a feature of construction without spending too much time. This type of drawing is based on the concept of a cube with three axes. See Fig. 2-2. One axis is a vertical line and the other two axes are sloped lines placed at 30° angles to a horizontal line. These axes and all lines drawn parallel to them are called isometric lines and are drawn in their true length. All measurements in these three directions are made without making any attempt to adjust the lengths to improve appearance. A box (a rectangular prism) ½″ × 1″ × 4″, Fig. 2-2, illustrates the value as well as the shortcomings of this type of drawing. The value lies in the ease with which it is drawn. The drawbacks are that there is no perspective as the lines go off into the distance, and the object is out of proportion because there is no foreshortening. This type of drawing is used to advantage to show details of construction. See Fig. 2-3. A line which is sloped presents a problem, but if the two ends of the line are located on isometric lines it can be drawn without difficulty. See Fig. 2-4. To draw the shed roof, dimensions A and B are measured on the vertical isometric lines and a line is drawn to connect the points. To draw the gable roof, dimensions A and B are measured and transferred to the isometric view. When the points are connected the gable end will be completed.

Oblique Drawing. One other type of pictorial representation is the oblique. One surface of the object is chosen by the draftsman

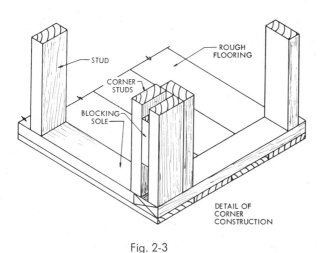

Fig. 2-3

An isometric drawing is used to show details of construction.

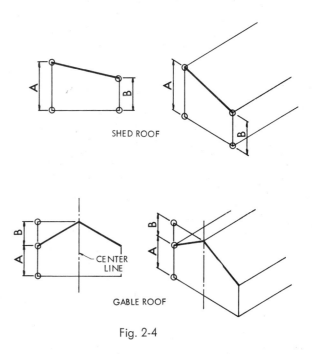

Fig. 2-4

Sloped lines (non-isometric) are drawn by locating points on isometric lines.

10   Building Trades Blueprint Reading

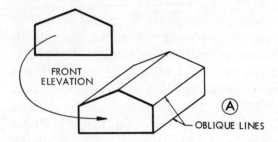

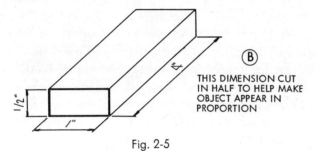

Fig. 2-5

Oblique Projection: This type of projection is useful in bringing about a three dimensional effect with a minimum of effort.

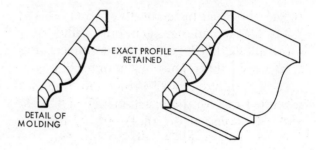

Fig. 2-6

Oblique projection is used to give illusion of depth.

to remain parallel to the imaginary plane he looks through to see the object. This surface is drawn in true shape using the actual dimensions. After drawing the shape of this surface he will draw parallel lines at an angle, usually 30° or 45° from each corner of the view. See Fig. 2-5A. In Fig. 2-5B the 4" dimension has been cut in half to keep the figure in proportion. Compare Fig. 2-5B with Fig. 2-2. Note that the isometric dimensions are actually 1" x 4" (to scale) while the oblique dimensions have been foreshortened to 1" x 2" (in scale).

An oblique pictorial drawing is a poor substitute for a photograph. However, this technique is useful in the study of blueprints because it creates an illusion of three dimensions with very little effort and may be used in conjunction with details shown on working drawings. Fig. 2-6 shows the profile of a molding such as may appear drawn in full size on a detail drawing. By adding a few parallel lines it is possible to visualize how the molding will look as used in a house.

Three View Drawing (Orthographic Projection). Pictorial drawings have great advantages for people not trained in blueprint reading and are a great help to those who want to learn this skill. However, the time spent in making pictorial drawings which would show the information necessary to build the house would be prohibitively expensive. The drawings would also be too complex to read. Three view drawings (or orthographic projection) are used almost universally in every field of architecture and engineering to make graphical representations of the jobs to be done. When looking at an object to be drawn the viewer must imagine that he is looking through a transparent plane. Lines are projected from every corner of the object to

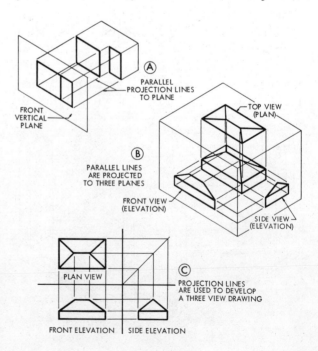

Fig. 2-7

Orthographic Projection: A three view drawing is visualized from a pictorial representation.

record the exact image on the plane. See Fig. 2-7A. Three planes are generally sufficient to show all of the details of most objects. Fig. 2-7B shows the planes as though they were part of a hinged box. Projection lines are extended to give the exact shapes of the top (the plan view of the roof), front, and side views of the house. The viewer imagines that he unfolds the planes into a flat position as shown in Fig. 2-7C. The projection lines between views serve to show how the various points and lines are related. It is important that this idea of the way views are related be understood from the beginning.

Fig. 2-8 shows how all of the views of the exterior of a house, including the roof plan, would be related if they were arranged to appear on the same sheet of paper. (The *floor plan* is drawn in order to give an idea of room arrangement but is not drawn in relation to the other views in this figure. The *plan view* in architectural drawing is a view seen by looking down vertically toward an object or, in other words, the view which is derived by projecting lines vertically upward to form an image on an imaginary horizontal plane. In Figs. 2-7B, 2-7C, 2-8B, and 2-8C, the plan view is that of the roof. Usually the plan views referred to are floor plan views.)

The architect when drawing a set of working drawings usually draws the floor plan first. Elevations are drawn on separate sheets for convenience. He is very careful to transfer all of the information about windows, doors, and

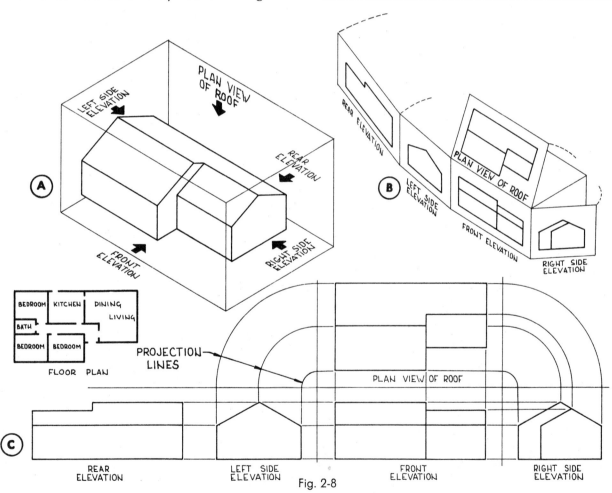

Fig. 2-8

Plan and elevation views are related by projection. Note how the projection lines connect the parts of one view to another. If the views in Fig. 2-8C were brought together following the projection lines, a three dimensional house would be formed as in Fig. 2-8A.

12 Building Trades Blueprint Reading

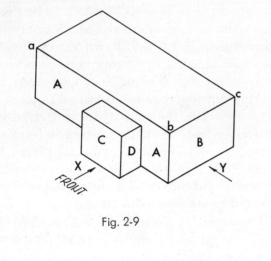

Fig. 2-9

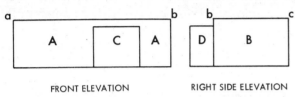

FRONT ELEVATION   RIGHT SIDE ELEVATION

Fig. 2-10

POINT OF VIEW

The front elevation of a house is generally toward the street. A person standing at Point X sees the front elevation. Fig. 2-9. When he moves to Point Y and stands back far enough, he will see the right side elevation. When the two elevations are drawn on the same sheet of paper they will appear as shown in Fig. 2-10. The roof is flat and is shown as line a-b in the front elevation and as line b-c in the side elevation.

Letters have been placed on the surfaces which are seen when looking from Points X and Y. Surfaces A and C are the only surfaces visible in the front elevation. Surfaces B and D are the only surfaces visible in the right side elevation. All of these surfaces are shown in their true shape and size in the elevation views because they are parallel to the plane through which the views are seen.

other exterior building features from the floor plan view to the elevation views. In effect, he is projecting lines from one view to another.

Fig. 2-8 shows the elevation views designated as Front, Left Side, Right Side, and Rear. This is accepted procedure particularly for stock plans which can be adapted for use on a lot facing in any direction. When a building is to be placed on a specific lot, it is more usual to give the compass directions: north, south, east, and west. The North Elevation is the elevation facing north (not the direction a person faces to see that side of the house).

Figs. 2-9 and 2-10 show the importance of relating the views correctly on the basis of the position of the observer. This was illustrated in Fig. 2-7A. The viewer takes different positions to see the various elevation views and must imagine lines drawn out from each point to strike a plane toward which he looks. Thus the image of the house drawn on the plane is the same shape and size (to scale) as that side of the house.

The ability to relate points in different views is a part of reading blueprints. Figs. 2-11 and 2-12 show how points which make up the corners of the roof are related in elevation drawings. When a series of points on a line such as a-b-c on the ridge are viewed in the front view, they are shown as one point because b and c are exactly behind a.

The idea of what is shown in its true shape and how true dimensions may be found is important. Surfaces and lines which are not parallel to the plane the viewer is looking through to see a particular elevation are not in their true length or shape. In the case of a roof it is necessary to take the length from one view and the dimensions along the slope from another view in order to construct the true shape. Sometimes this information is required.

Figs. 2-13 and 2-14 provide additional practice in identifying planes and lines in the elevation views of a building. By studying the elevation views, the relationship of the masses making up the two parts of the building can be determined. The relationship of one line to another in height above the grade (ground level) is determined by how the various lines are related by projection in the views shown.

Reading Drawings 13

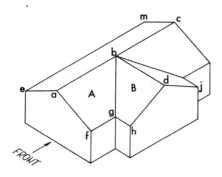

Fig. 2-11

USING SEVERAL VIEWS TO FIND INFORMATION

It is often necessary to look at several elevations in order to obtain all of the required information. Letters have been used to designate some of the points of the roof and two of the roof surfaces to bring out this idea. Study Fig. 2-11 to see how most of the points determining the ends of lines have been assigned letters. Looking directly toward the front of the house as shown by the arrow, notice how:

Point m lines up behind Point e.
Point b and c line up behind Point a.
Point g lines up behind Point f.
Point j lines up behind Point h.

Looking from the right side to see the right side elevation, notice how:

Point e lines up behind Point f.
Point g lines up behind Point h.
Point b lines up behind Point d.

Referring to Fig. 2-12, notice that the main ridge of the house is shown in the right side elevation as line a-c. To find the true length of the ridge of the projecting ell it is necessary to find length b-d in the front elevation. That these two ridges are the same height above the grade (ground level) is shown by the fact that they intersect at Point b. See Fig. 2-11 and Fig. 2-12.

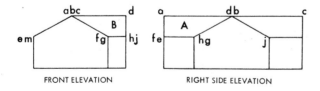

Fig. 2-12

The eaves of the house are all at the same level because all of the Points e, f, g, h, and j are on the same line by projection in the front elevation and in the right side elevation.

The true length of line g-h is found in the front elevation. The line which shows the slope of roof A in its true length is line a-f seen in the front elevation. The line which shows the slope of roof B in its true length is line d-h seen in the right side elevation.

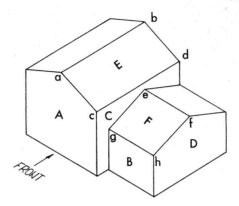

Fig. 2-13

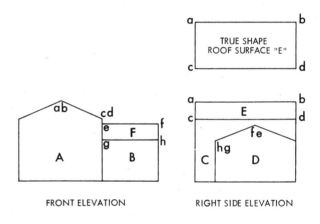

Fig. 2-14

LOCATING SURFACE, LINE, AND LEVEL INFORMATION IN ELEVATION VIEWS

Fig. 2-13 represents a tri-level house. Capital letters are used to designate some of the surfaces, and small letters are used to designate some of the points. These will aid in understanding the elevations in Fig. 2-14. Surfaces A and B are seen in the front elevation in their true shape and size. Surfaces C and D are seen in the right side elevation in their true shape and size. Roof Surface E is seen in the right side elevation but it is not in its true shape because it is sloped away from the plane through which the view is observed. (The true length of Surface E is shown in the right side elevation, however, by line a-b or line c-d.) In order to show the true shape and size of Surface E, another figure would have to be drawn. The true length dimension may be taken from the right side elevation (line a-b or c-d) and the true width dimension may be taken from the front elevation (line a-c). With these dimensions, a rectangle may be drawn which will show the true shape and size of roof Surface E. Roof Surface F is not in its true shape either. What lines from the elevations would be used to draw a rectangle showing the true shape and size of Surface F?

14 Building Trades Blueprint Reading

TEST DRAWING NO. 1 (See pages 15-16)

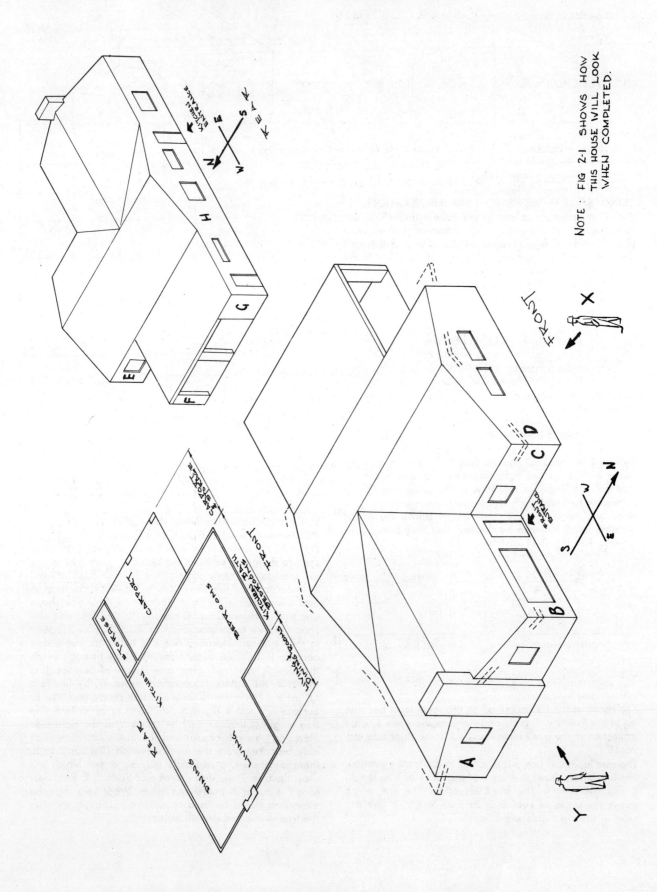

NOTE: FIG 2-1 SHOWS HOW THIS HOUSE WILL LOOK WHEN COMPLETED.

Reading Drawings   15

Trade Competency Test No. 1
(Chapter 2: Based on Test Drawings Nos. 1, 1A and 1B)

STUDENT'S NAME   INSTRUCTOR'S NAME

MULTIPLE CHOICE TEST: Circle the correct answer or answers.

		ANSWER	SCORE
Example: How many windows will be shown on the South Elevation drawing?		1 2 3 ④ 5	
1. What type of pictorial drawing is used on Test Drawing 1?	1.	Perspective Isometric Oblique	
2. A person standing at position X would see which elevation of the house? (North, South, East or West Elevation.)	2.	N S E W	
3. A person standing at position Y would see which elevation of the house?	3.	N S E W	
4. The kitchen entrance will be shown on which elevation drawing?	4.	N S E W	
5. How many doors will be shown on the South Elevation drawing?	5.	1 2 3 4 5	
6. How many windows will appear on the East Elevation drawing?	6.	1 2 3 4 5	
7. How many windows will appear on the West Elevation drawing?	7.	1 2 3 4 5	
8. Which three elevations will show the most information about the chimney?	8.	N S E W	
9. Which two elevations will show the most information about the car port?	9.	N S E W	
10. How many exterior doors are shown on the views in Test Drawing No. 1?	10.	1 2 3 4 5	
11. The sloped gable end of the roof over the bedrooms will show in which elevation drawing?	11.	N S E W	
12. Which wall surfaces will be seen on the drawing of the North Elevation?	12.	A B C D E F G H	
13. Which wall surface will be seen on the drawing of the South Elevation?	13.	A B C D E F G H	
14. Which wall surfaces will be seen on the drawing of the West Elevation?	14.	A B C D E F G H	
15. Which wall surfaces will be seen on the drawing of the East Elevation?	15.	A B C D E F G H	

Tear Off Here

(Cont.)

16 Building Trades Blueprint Reading

Trade Competency Test No. 1 (Cont.)

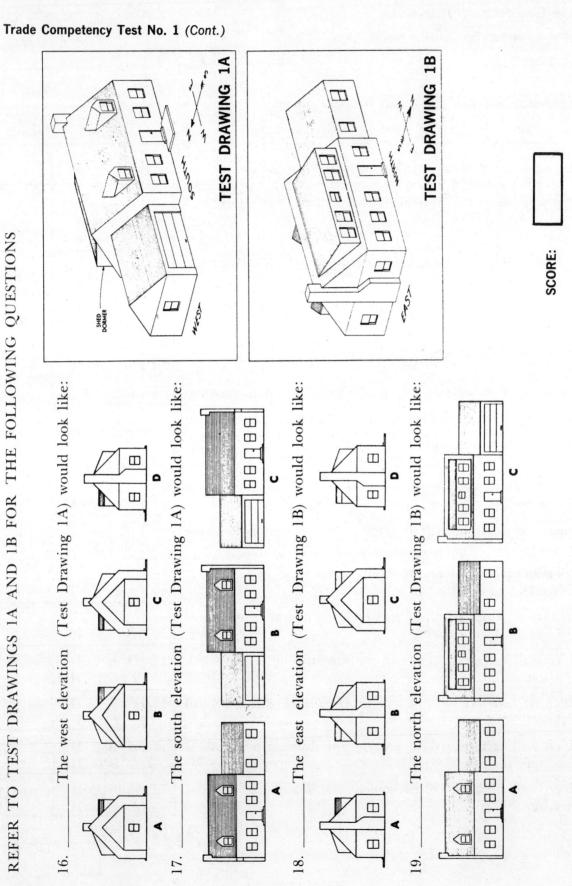

REFER TO TEST DRAWINGS 1A AND 1B FOR THE FOLLOWING QUESTIONS

16. _____ The west elevation (Test Drawing 1A) would look like:

17. _____ The south elevation (Test Drawing 1A) would look like:

18. _____ The east elevation (Test Drawing 1B) would look like:

19. _____ The north elevation (Test Drawing 1B) would look like:

SCORE:

Tear Off Here

Chapter 3

Reading Elevation Drawings

An elevation drawing is an orthographic projection showing the exterior view of one side of a building. (An elevation drawing may be made of an interior wall or feature of a building, but for the purpose of this chapter the elevations are all considered to be exterior views.) The four elevations, each showing a side of the structure, are part of the working drawings prepared by the architect and serve to indicate what the building will look like when it is completed. Their function is to show the design of the house, where the openings are to be placed, what materials are to be used, etc.

It is important to understand how the blueprints of a set of working drawings are related to each other and why it is often necessary to refer to several of them in order to find all of the information on one subject. The original drawings are carefully drawn by the architect so that foundation plan, floor plans, and the elevations exactly match regarding the location of windows, doors, and other details. For instance, to find out about a particular window it is necessary to look at the blueprint of the floor plan to find out the exact dimension from the corner of the house to the window. An elevation view shows what the window looks like and where it is located vertically in the wall. The size of the window may appear on the elevation drawing but it may also be found on a window schedule on some other blueprint. Ample opportunity will be given to work with this relationship between the floor plan and elevation blueprints in a later chapter.

The architect shows much of the information on elevation drawings by using symbols and abbreviations. The reason for using these shortcuts is due to the fact that the drawings are usually made at 1/48th the size of the building ($1/4'' = 1'\text{-}0''$). To show all of the lines which appear on the finished building and to spell out each designation would be impossible. A sort of shorthand of lines and symbols has been adopted by architects so that plans may show the necessary information to the builder without becoming so complex that they would be difficult to read. See Fig. 3-1. The same is true of abbrevia-

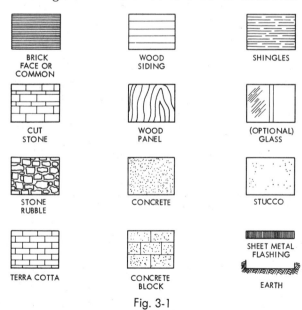

Fig. 3-1

SYMBOLS OF MATERIAL IN ELEVATION
Symbols for materials reduce the need for notations on drawings.

17

tions. Architects have devised abbreviations which have become accepted in the industry and which are used to designate the many things which appear on the blueprints. The readability of the blueprint is preserved because there is some space left not covered by notations. See Fig. 3-2.

A set of specifications accompanies each set of blueprints. The specifications give general information about the legal aspects of responsibility, guarantees of performance, etc. Following the general information sections, the specifications describe the responsibility of each subcontractor as to what work is to be done and what materials are to be used. This subject is covered in *Building Trades Blueprint Reading, Part 2*.

Information to be Found on Elevation Drawings. This is a broad checklist. Mark the terms which are new to you for further study.

Design of the Building.
General shape. Location of offsets, ells, patios, steps, porches, bays, dormers, chimneys, etc.
Information on footings and foundation.
The Roof.
Type and slope of roof.
Roofing material.
Vents, gravel stop, projection of eaves.
Openings.
Windows — types, sizes, swing, location in wall.
Doors — types, sizes, location in wall.
Dimensions.
Dimension from established grade to finished first floor level.
Dimension from established grade to finished basement floor level.
Floor to floor heights.
Heights of special windows above floor.
Dimension from ridge to top of chimney.
Exterior Finish—Material.
Types of wood or composition siding, shingles, etc.
Concrete, concrete block, brick, stone, stucco, etc.

Exterior Finish—Trim.
Decorative treatment at:
Windows.
Entrance doorways.
Columns, posts, ballustrades.
Cornices.
Miscellaneous Details.
Electrical fixtures and utility outlets.
Gutters and downspouts.
Flashing and waterproofing.

Design of the Building. A house is generally designed from the inside out. In other words, the floor plans are considered first because they will determine the arrangement and size of the various space divisions. The exterior of the house is no less important because it reflects the good planning used inside and makes the house a thing of beauty outside. Often an architect is expected to design a home in a particular style such as the Cape Cod, Georgian, or contemporary ranch. The details of design, such as the general proportions, the roof, type of windows and trim, must be in keeping with the style. As the student reads the blueprints he must learn how to distinguish the parts of the building by looking at the elevations. Whenever a change is shown in a wall or roof line it indicates some modification of the rectangular shape. After studying several sets of plans, he should be able to see from the elevations, with only a quick reference to the first floor plan, whether the building is "L," "U" or "T" shaped or has an enclosed court. Other elements of the building such as bays, dormers, and chimneys should be studied by looking at two related elevations so that the front and side views may be observed.

The part of the building under the established grade (ground level) is shown with hidden (dashed) lines. This includes footings (the foundation base), foundation walls, and windows and areaways below grade related to that particular side of the house. There is no other blueprint in the set where this information could be shown as suitably.

Reading Elevation Drawings

ABBREVIATIONS COMMONLY USED ON ELEVATIONS

Aluminum . . . . AL	Insulating (Insulated) . . . . INS
Asbestos . . . . ASB	Length L
Asphalt . . . . ASPH	Length Over All . . . . LOA
Basement . . . . BSMT	Level . . . . LEV
Beveled . . . . BVL or BEV	Light . . . . LT
Brick . . . . BRK	Line L
Building . . . . BLDG	Lining . . . . LN
Cast Iron CI	Long . . . . LG
Ceiling . . . . CLG	Louver . . . . LVR or LV
Cement . . . . CEM	Low Point . . . . L PT or LP
Center . . . . CTR	Masonry Opening . . . . MO
Center Line . . . . ℄ or CL	Metal . . . . MET. or M
Clear . . . . CLR	Moulding . . . . MLD or MLDG
Column . . . . COL	Mullion . . . . MULL
Concrete . . . . CONC	North . . . . N
Concrete Block . . . . CONC BLK	Number . . . . NO. or #
Concrete Masonry Unit . . . . CMU	Opening . . . . OPG or OPNG
Copper . . . . CPR or COP	Outlet . . . . OUT
Corner . . . . COR	Outside Diameter . . . . OD
Detail . . . . DTL or DET	Overhead . . . . OH or OVHD
Diameter . . . . DIAM or DIA or ⌀	Panel . . . . PNL
Dimension . . . . DIM	Perpendicular . . . . PERP
Ditto . . . . DO.	Plate Glass . . . . PG or PL GL
Divided . . . . DIV	Plate Height . . . . PL HT
Door . . . . DR	Radius . . . . RAD or R
Double-Hung Window . . . . DHW	Revision . . . . REV
Down . . . . DN or D	Riser . . . . R
Downspout . . . . DS	Roof . . . . RF
Drawing . . . . DWG	Roof Drain . . . . RD
Drip Cap . . . . DC	Roofing . . . . RFG
Each . . . . EA	Rough . . . . RGH
East . . . . E	Saddle . . . . SDL or S
Elevation . . . . EL	Scale . . . . SC
Entrance . . . . ENT	Schedule . . . . SCH
Excavate . . . . EXCA or EXC	Section . . . . SEC or SECT
Exterior . . . . EXT	Sheathing . . . . SHTH or SHTHG
Finish . . . . FIN,	Sheet . . . . SHT or SH
Fixed Window . . . . FX WDW	Shiplap . . . . SHLP
Flashing . . . . FLG or FL	Siding . . . . SDG
Floor . . . . FLR or FL	South . . . . S
Foot or Feet . . . . ' or FT	Specifications . . . . SPEC
Footing . . . . FTG	Square . . . . SQ or □
Foundation . . . . FND	Square Inch . . . . SQ IN. or □"
Full Size . . . . FS	Stainless Steel . . . . SST
Galvanized . . . . GV or GALV	Steel . . . . ST or STL
Galvanized Iron . . . . GI	Stone . . . . STN
Galvanized Steel . . . . GS	Terra Cotta . . . . TC
Gage . . . . GA	Thick or Thickness . . . . THK or T
Glass . . . . GL	Typical . . . . TYP
Glass Block . . . . GLB or GL BL	Vertical . . . . VERT
Grade . . . . GD or GR	Waterproofing . . . . WP
Grade Line . . . . GL	West . . . . W
Height . . . . HT	Width . . . . W or WTH
High Point . . . . H PT	Window . . . . WIN or WDW
Horizontal . . . . HOR	Wire Glass . . . . WG or W GL
Hose Bibb . . . . HB	Wood . . . . WD
Inch or Inches . . . . " or IN.	Wrought Iron . . . . WI

Fig. 3-2

Architects and engineers find it necessary to use abbreviations in order to save time and conserve space on working drawings. Most abbreviations are accepted through common usage. The American Institute for Architects in a report entitled "A Uniform System for Working Drawings" (1974) reviewed the problem and suggested some changes which are included in the above list. Where there are two abbreviations, the AIA form is given first. Only capital letters are to be used for abbreviations. Abbreviations which make an actual word, such as IN., are usually followed by a period.

Self-Check Quiz No. 3-A Elevation Abbreviations

(Based on Fig. 3-2)

COMPLETION TEST: Study the list of elevation abbreviations given on Fig. 3-2. In the space provided fill in the word for the abbreviated word. Only fill in the abbreviated word or words. *Answers are given in the appendix.*

Example: ASPH SHINGLES   <u>ASPHALT</u>

1. COMMON BRK WALL   1. _____
2. BLDG LINE   2. _____
3. CI COLUMN   3. _____
4. CEM CAP   4. _____
5. CLG LINE   5. _____
6. CONCRETE COL   6. _____
7. CONC WALK   7. _____
8. CPR DOWNSPOUT   8. _____
9. COR OF STUD   9. _____
10. 24'-0" DIAM TANK   10. _____
11. APPROXIMATE DIM   11. _____
12. SASH DO.   12. _____
13. SASH 32 × 24 DIV   13. _____
14. 12 LIGHT DHW   14. _____
15. 14 RISERS DN   15. _____
16. COPPER DS   16. _____
17. REFER TO SECTION DWG   17. _____
18. FRONT EL   18. _____
19. ENT DETAIL   19. _____
20. EXT FINISH   20. _____
21. FIN. FLOOR   21. _____
22. CHIMNEY FLG   22. _____
23. CONC FLR   23. _____
24. 32 FT   24. _____
25. GI GUTTER   25. _____
26. GL BLOCK   26. _____
27. 4'-0" ABOVE GD   27. _____
28. 3'-0" ABOVE SILL HT   28. _____
29. H PT OF LOT   29. _____
30. 17 IN.   30. _____
31. L OF OPENING   31. _____
32. 6 LT WINDOWS   32. _____
33. BLDG L   33. _____
34. PIPE 37" LG   34. _____
35. LPT OF LOT   35. _____
36. MET. THRESHOLD   36. _____
37. WOOD MLD   37. _____
38. HORIZONTAL MULL   38. _____
39. STACK 36" OD   39. _____
40. 3'-0" × 6'-8" PNL DR   40. _____
41. PG STORE FRONT   41. _____
42. MEASURE FROM PL HT LINE   42. _____
43. ARCH HEAD 3'-0"R   43. _____
44. ASPHALT RFG   44. _____
45. RGH TROWELLED CEMENT   45. _____
46. SC ¼" = 1'-0"   46. _____
47. INSULATED SHTH   47. _____
48. ¾" PLYWOOD SHT   48. _____
49. SHLP SIDING   49. _____
50. REFER TO SPEC   50. _____

Reading Elevation Drawings

The Roof. The roof of the house serves to shelter the structure from the elements but also serves an artistic purpose. In reading blueprints of elevation drawings for information on roofs this becomes apparent. The various types of roofs used are flat, shed, gable, hip, gambrel, mansard, or a combination or modification of these types. See Figs. 3-3 and 3-4. The slope of the roof is quite important because it is directly related to snow load, rain dispersal, and appearance. The gable roof in Fig. 3-3 shows a slope triangle with 5 on the vertical leg and 12 on the horizontal leg. This means that for every foot of horizontal measurement there is a vertical rise of 5 inches. The amount of projection at the eaves is shown either on the blueprints of the elevations or on a blueprint showing a vertical section through the wall. (Projection is measured horizontally; overhang is measured parallel to the roof slope.) When the roof is very complex the architect may draw a roof plan. Generally however the person reading the plan must determine the shape of the roof by comparing several elevations.

The roofing material and the manner of laying it is usually shown unless this information is covered in detail in the specifications.

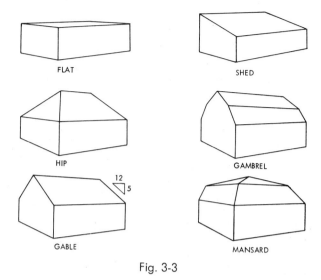

Fig. 3-3

SIX BASIC ROOF TYPES
Basic roofs follow traditional styles. Other shapes are formed by modifying and combining these types.

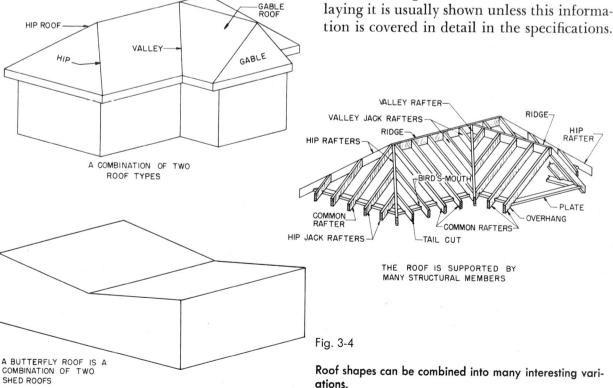

Fig. 3-4

Roof shapes can be combined into many interesting variations.

Openings. Blueprints of elevation drawings show a great deal about windows, doors, and other openings. Reference must be made to the blueprints of floor plans to find horizontal locating dimensions.

The elevations show the windows as they appear in the wall in their exact location. The window types most frequently used are fixed sash, double hung, horizontal sliding (gliding), casement, awning, and hopper. The elevation views show how the sash is divided and where the hinges are located on a sash which swings in or out. The apex of a triangle drawn on a window points to the side with the hinges. See Fig. 3-5. The size of the light (pane of glass) may also be shown. If a sash of a window contained a designation 28/24, this would mean that the glass size is 28 inches wide and 24 inches high. Fig. 3-5 shows typical window symbols.

Doors are indicated on blueprints of elevation drawings in their proper location and their particular style is shown. The symbol for a flush door is a plain rectangle unless it has a light (glass). A panel door is drawn to show the panel and light arrangement. See Fig. 3-5. The lock and hinges are usually omitted. The floor plan shows how the door swings. The builder refers to the floor plan to find out on which side of the door to install the lock and on which side to install the hinges.

Other openings, such as screened vents, louvers, etc., will be drawn in place and designated with whatever notes are deemed necessary.

Dimensions. Several important functions of elevation views are to establish floor levels, to relate these levels to the finished grade, and to supply dimensions which would not appear elsewhere on the set of blueprints. The floor levels drawn on the elevations help the builder visualize the relationship between windows, and floor and ceiling levels. This information may also appear on a blueprint which shows a vertical wall section.

Exterior Finish — Materials. The elevation views show by means of symbols and notations what type of material is used on the outside of the house. Siding of several shapes, widths, and materials is often required. Plywood sheets serve as exterior wall covering and are used in a variety of ways. Shingles of wood, asbestos cement, and asphalt composition, as well as stucco, give several interesting effects. Masonry of brick, stone, or concrete blocks serve structural purposes and serve as wall facing. A veneer of brick or stone has proved to be very satisfactory in many regions. (See Fig. 3-1 for symbols.)

Exterior Finish — Trim. The elevation views also show decorative features designed to enhance the appearance of the building such as an entrance doorway, a ballustrade, columns or posts, decorative trim around windows, and the cornice at the eaves.

Miscellaneous Details. A number of other things may be shown on the blueprints of the elevations to alert the builder to the location of specific items. Electric utility outlets and fixtures are indicated at their proper places and at the approximate height desired. Gutters at the eaves and downspouts (sometimes called leaders) are drawn in place, and flashing (metal strips) is shown at chimneys and other places where water might find its way between masonry and wood members. Hose bibbs may be shown but are more usually indicated on the plan views only. Fig. 3-6 illustrates many of these details.

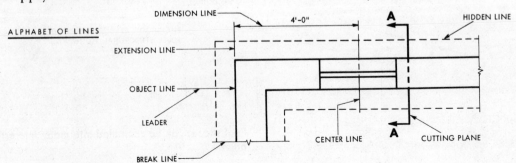

Reading Elevation Drawings 23

SYMBOLS USED FOR WINDOWS AND DOORS ON ELEVATION DRAWING

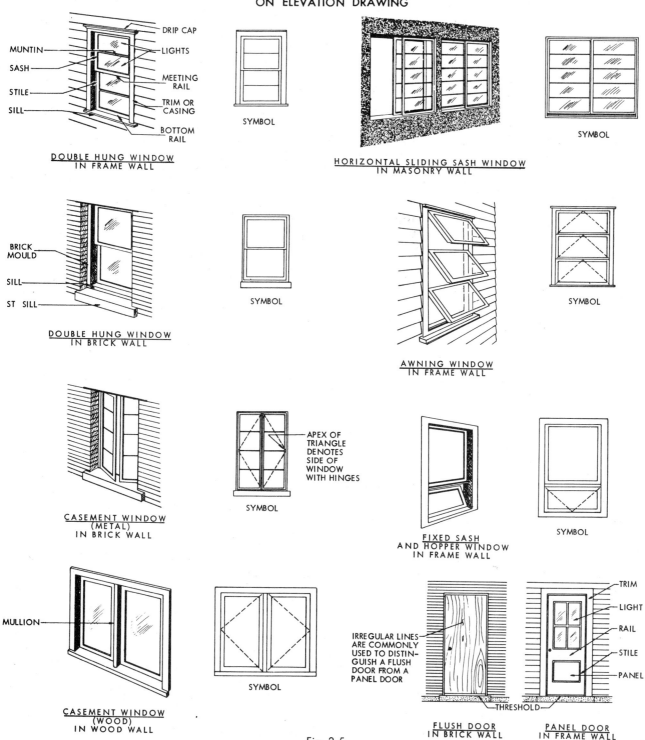

Fig. 3-5

WINDOW AND DOOR SYMBOLS

Many lines are omitted when symbols are drawn. Wood windows have wider sash parts than metal windows. Windows in frame walls have wood trim. Windows in masonry walls have a narrow brick mold. The side which has the hinges on swinging windows is indicated by the apex (or point) of a dashed triangle. (Note the swing symbol for a casement window.) A stationary window with no provision for opening is called a "fixed sash window." Locks and hinges on doors are not generally shown. It is necessary to refer to the floor plan to see the swing of the doors.

24 *Building Trades Blueprint Reading*

Self-Check Quiz No. 3-B
(Based on Figs. 3-3 and 3-4)

COMPLETION TEST: Sketch the approximate shape of the FRONT and RIGHT SIDE ELEVATIONS in the space provided. *Answers are given in the appendix.*

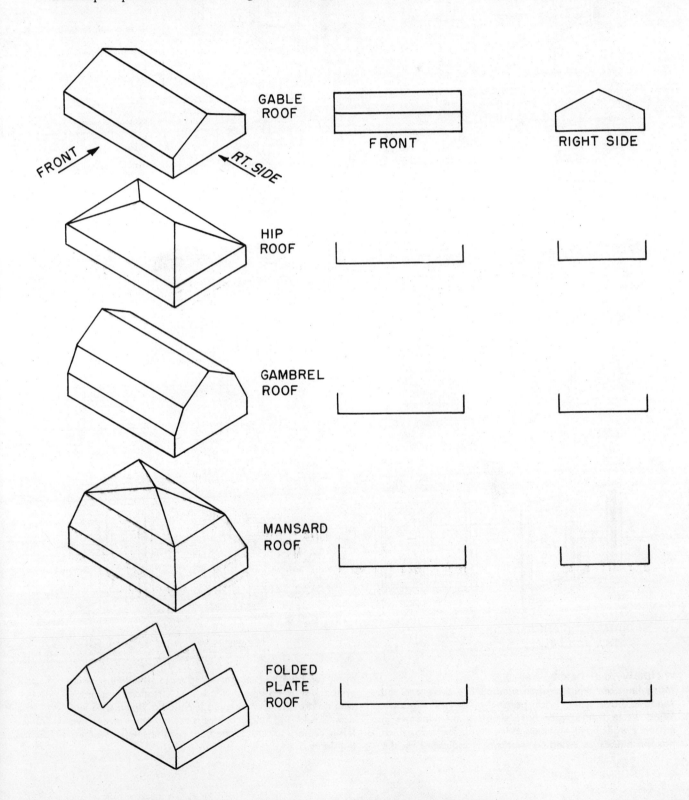

Reading Elevation Drawings 25

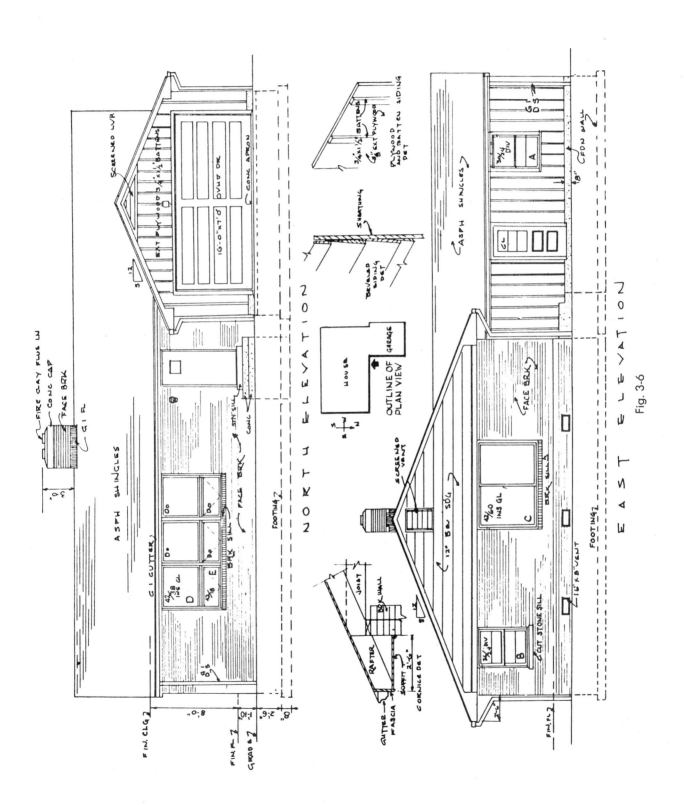

Fig. 3-6

26 Building Trades Blueprint Reading

Self-Check Quiz No. 3-C on Elevation Drawings
(Based on Fig. 3-6)

The house chosen for the first blueprint reading exercise is typical of houses which have been erected in many parts of the country. It shows a number of the things discussed in this chapter using symbols, abbreviations, and dimensions as they might appear on working drawings. Answer the questions in the spaces provided. *Answers are given in the appendix.*

1. Does the house with the garage form a T shape, L shape, or U shape? _____
2. What type of roof is used? _____
3. What material is used for roofing? _____
4. What is the slope of the roof? _____
5. What is used to ventilate the space under the roof of the house? _____ of the garage? _____
6. How far does the cornice project beyond the face of the brick wall of the house? _____
7. What type of front entrance door is shown? _____
8. What kind of a door is used at the side of the garage? _____
9. What is the size of the garage overhead door? _____
10. What type is window "A"? _____
11. What do the designations 30/24 DIV mean? _____
12. What type is window "B"? _____
13. What type is window "C"? _____
14. What does "INS GL" mean? _____
15. What type is window "D"? _____
16. What type is window "E"? _____
17. Are the hinges at the top or bottom of window "E"? _____
18. Give the following dimensions:
    Grade to bottom of footing _____
    Grade to finished first floor _____
    Finished first floor to finished ceiling _____
    Ridge to top of chimney flue _____
19. What type of siding is used on the garage? _____
20. Since the garage siding must not extend down to the grade, how much space must be allowed according to these elevations? _____
21. What material is used on the outside surface of the house? _____
22. What kind of a chimney cap is used? _____
23. What is used to ventilate the crawl space? _____
24. Where is flashing indicated? _____
25. How many exterior lighting fixtures are shown? _____

Reading Elevation Drawings 27

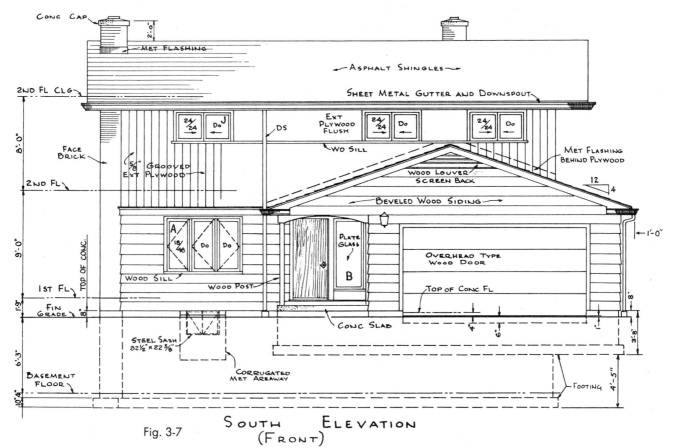

Fig. 3-7 SOUTH ELEVATION (FRONT)

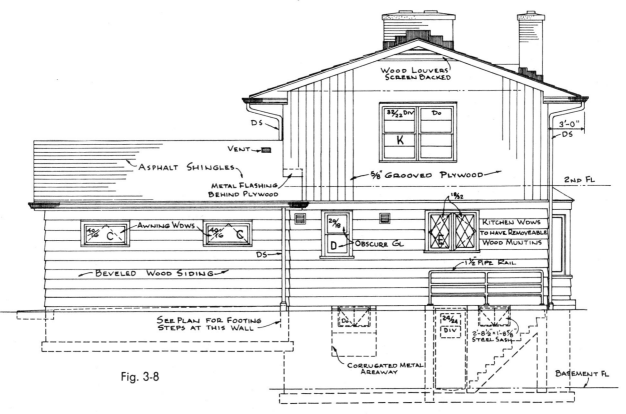

Fig. 3-8

EAST SIDE ELEVATION

(See Trade Competency Test No. 2, pages 29-30)

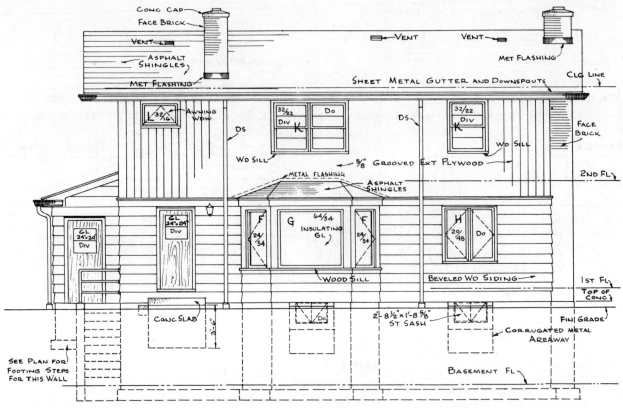

Fig. 3-9

NORTH ELEVATION

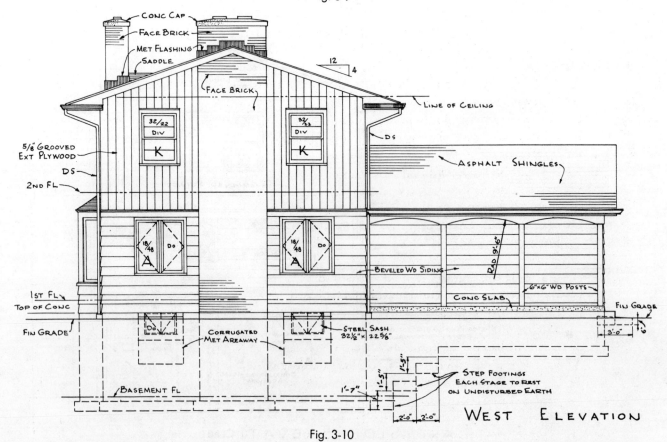

Fig. 3-10

WEST ELEVATION

(See Trade Competency Test No. 2, pages 29-30)

Trade Competency Test No. 2
(Chapter 3: Based on Figs. 3-7, 3-8, 3-9, and 3-10)

STUDENT'S NAME   INSTRUCTOR'S NAME

MULTIPLE CHOICE TEST: The following statements are incomplete. Choose the phrase which best completes the statement and put the corresponding letter in the proper space in the answer column.

Example:

	Answer	Score

One type of exterior finish used on the house is
  A. boards and battens.   C. beveled wood siding.
  B. cedar shingles.   D. mineral fiber.
	C	

1. The house can best be described as
  A. a one-story house with attic.
  B. a two-story rectangular house with attached garage.
  C. a modified ranch house.
  D. a two-story L-shaped building.
1.		

2. Looking at the elevation drawings, it is possible to see
  A. garage walls in E and S elevations only.
  B. gutters and downspouts on the rear of the house on the N, E, and W elevations only.
  C. the main chimney in the W, N, and S elevations only.
  D. the bay in the S elevation only.
2.		

3. Which of the following statements is true?
  A. The roof on the house has a slope of five inches per foot.
  B. The garage roof is a hip roof.
  C. Eaves on house and garage have the same projection.
  D. The roof on the house is a gable roof with wide eaves.
3.		

4. The garage
  A. roof extends over a walk to the front entrance.
  B. concrete floor is level.
  C. is ventilated by doors and windows only.
  D. apron extends 3'-0" in front of the front wall.
4.		

5. The bay
  A. has a roof of sheet metal.
  B. has the same slope as the main roof.
  C. may be seen in only the N elevation.
  D. overhanging gutter and downspout are omitted.
5.		

6. The dimension from the
  A. grade to the top of the concrete is 8".
  B. grade to the second floor level is 9'-8".
  C. first floor level to the basement level is 8'-4".
  D. bottom of the garage footing to the top of the foundation wall is 4'-2".
6.		

7. Which of the following statements is true?
  A. There is a basement under the house and garage.
  B. There is a footing under the foundation for the slab at the rear door to the house.
  C. There is an exterior entrance to the basement.
  D. The stepped footings are under the house foundation.
7.		

(Cont.)

Trade Competency Test No. 2 *(Cont.)*

8. Basement windows are
    A. completely below grade.
    B. seven in number with six areaways.
    C. all the same size.
    D. wood sash with wood frames.
9. To ventilate the space under the roof
    A. louvers are placed on both gable ends of the house.
    B. louvers are placed on both ends of the garage.
    C. vents are placed on both sides of the ridge of the house.
    D. a louver in the gable end and a vent near the ridge have been used in the garage.
10. Which of the following statements is true?
    A. Flashing is shown at chimneys to add to appearance.
    B. Flashing is shown over doorways.
    C. Flashing over the garage roof helps keep out water.
    D. The saddle at the chimney may be seen in all elevations.
11. Which of the following statements is true?
    A. The rear garage door is shown on the N elevation.
    B. A light is indicated over the rear door to the house.
    C. There is a light indicated at the front entrance doorway.
    D. Hose bib locations are shown.
12. Regarding exterior stairs to the basement:
    A. the stairs descend toward the north.
    B. the stairs are partially above grade.
    C. the stairs are indicated in three elevations.
    D. a pipe railing encloses two sides of the opening.
13. Excluding the garage door, information about doors
    A. indicates that all doors are panel doors.
    B. indicates that all doors are flush doors.
    C. indicates that all doors are flush doors with glass.
    D. indicates the sides of the doors with the hinges.

On the drawings on pages 27 and 28 each window or pair of windows is marked with a letter. In questions 14 through 20 place a check in *one* box under each letter to show what *type* of window the letter indicates. "K" windows are double hung, so place the check under "K," and level with number 18.

First floor windows designated:
14. Double hung window   14.
15. Fixed sash window   15.
16. Casement window   16.
17. Awning window   17.

Choose correct answer and place checks in corresponding squares.

Second floor windows designated:
18. Double hung window   18.
19. Horizontal sliding window 19.
20. Awning window   20.

SCORE:

Chapter 4

Reading Floor Plans

Importance of Floor Plans. The building tradesman finds it to his advantage to be able to read blueprints of floor plans very early in his job experience. When he acquires this ability, he becomes part of a team representing many skills working out a series of construction problems together. He must know how to "take off" dimensions accurately so that all of the partitions, windows, and doors are located according to the blueprints. He must learn to recognize the symbols representing material, equipment and fixtures, and to interpret abbreviations and understand notations. He should know about information intended for trades other than his own. Almost every trade is inter-related to some extent, and all tradesmen must follow the prints carefully if the work is to progress smoothly.

This chapter is devoted to learning how to read floor plans in their simplest form. When an experienced builder gets a new set of blueprints to study, he spends some time becoming familiar with the layout, paying little or no attention to the maze of information usually found on them. The broad aspects of shape, size and relationship of rooms, and the use of auxiliary space such as hallways, stairs, and closets are of primary importance. Fig. 4-1 shows a simplified layout of the working drawing shown on Test Drawing No. 4 in Chapter 5, page 60. (It is introduced at this point not for study but to show the difference between a simplified floor plan and the actual working drawing.) The process of learning to visualize how a house is laid out requires this type of simplification. Sometimes a blueprint is so complex that it is advisable to place a piece of tracing paper over it and to trace the walls, partitions, and openings, omitting all other information. This procedure will provide a sketch showing the relationship between areas which is comparatively easy to

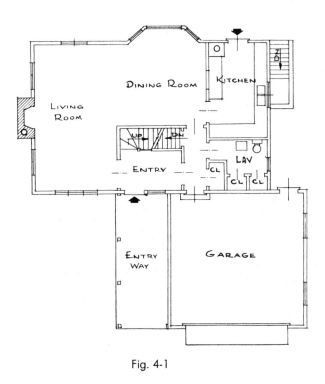

Fig. 4-1

A simplified plan view is easy to read. CL is an abbreviation for "closet." (Test Drawing No. 4 in Chapter 5 is the same plan in working drawing form.)

32 Building Trades Blueprint Reading

read. Some imagination must be used in studying floor plans. The student should try to visualize what he would see as he entered the front door of the house. Would he see a stairway perhaps, or a living room with a fireplace? As his mind takes him from room to room, where would the windows be, the closets, the cabinets, etc.?

In planning a house the greatest emphasis is put on the floor plans. An architect generally feels that if the floor plans are good, the elevations will be relatively easy to work out. (This is not always true because special problems may arise anywhere.) In learning to read blueprints the greatest emphasis, therefore, should be made on studying floor plans because the floor plans are of primary importance.

Basic Ideas in Reading Floor Plans. The following things are basic to reading blueprints of plan views.

1. Floor plan views are drawn to exact scale. The rooms, hallways, cabinets, stairs, etc., are drawn so that they are in correct relationship to one another. Chapter 6 will discuss how to read dimensions and how scales are used to show the various parts of working drawings to the best advantage.

2. The floor plan views and elevations are drawn to the same scale and are exactly related to each other. Windows which appear on the floor plan views are the same size and the same distance from the building corners as they are on the elevation views. By referring back and forth from floor plan views to elevation views it should become relatively easy to visualize the whole house.

3. Floor plan views are related to each other. Structural provisions are made so that the load of floors and partitions are transferred to supporting members or partitions immediately below. Stairs are designed so that they start on one floor and end in the right place on the floor above or below. Provision is made so that heating ducts may start at the furnace, pass through first floor partitions, and end at registers in the desired location in second floor partitions.

It is generally considered good practice not to show the same information in two

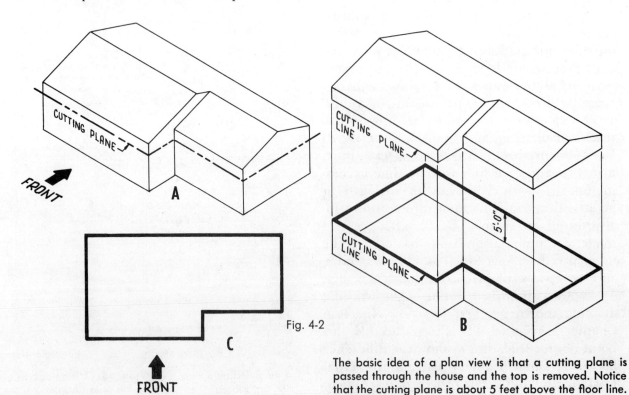

Fig. 4-2

The basic idea of a plan view is that a cutting plane is passed through the house and the top is removed. Notice that the cutting plane is about 5 feet above the floor line. (See "B".) Also observe that the front of the building is facing toward the bottom of the sheet. (See "C".)

places. For instance, a stairway which ascends from the first to second floor is never shown completely on the first or second floor plans. The first floor plan will show the exact location of the bottom riser and will show a few treads, terminating the stairway on a break line. (The riser is the vertical portion of a step and the tread is the horizontal portion.) The second floor plan shows the top riser location and a few of the descending stair treads terminating on a break line also. (Fig. 4-1 illustrates stairs leading to the second floor and to the basement.) A window which appears on a stair landing and could be shown on both the first floor plan and the second floor plan is shown on only one or the other plan.

4. Floor plan views are generally drawn so that the front view is toward the bottom of the sheet. This follows the fundamental concept of three view drawing. See Figs. 4-1 and 4-2. When a house is long and narrow, such as one which must fit on a narrow city lot, the front is usually placed facing toward the right edge of the sheet.

5. The floor plan view is a horizontal slice taken through the house at between four to five feet above the floor. The plane cuts through upper kitchen cabinets, medicine chests in bathrooms, and the upper sash of windows. However, this rule is modified to cut through fireplaces or other building parts which need to be shown on the plan at a lower level. A significant departure from the idea of looking down at a slice taken through a house to see the floor plan is followed in noting structural members. A designation "2 × 10 Joists over" on a first floor plan means that 2 × 10 inch joists (supporting members) are *overhead* supporting the second floor.

Fig. 4-2 shows how a cutting plane is passed through a house in order to show the interior. Actually, a floor plan view is a horizontal section view.

Reading Floor Plans. Four houses, Figs. 4-3, 4-5, 4-13, and Test Drawing No. 3, have been chosen for study in this chapter to bring out the idea of reading floor plans in their simplest form. All of the dimensions and detailed information have been left off, and symbols for materials of construction, windows, doors, etc., have been greatly modified. The purpose is to make it possible to see room shapes and relationships without distracting detail.

Pictorial representations of three of the houses will help you to visualize a flat two-dimensional plan by introducing a three-dimensional concept. Study the floor plan views using the pictorial drawings freely so that you may understand all of the parts of the floor plan. Remember, however, pictorial views are rarely available, and you must gradually arrive at a point where you can understand all about a house by studying the blueprints of the working drawings without these aids.

One Story House. Fig. 4-3 is a sketch floor plan of a simple one floor dwelling. Fig. 4-4 is the same house in isometric, cut to show the layout of partitions and location of openings. Imagine yourself entering the front door and walking through the living room to the hall and then to the bathroom and bedrooms. Try to get an idea of the size and proportions of each room. Notice how the windows are

Fig. 4-3

A sketch plan of a house shows the room layout, location of windows and doors, plumbing fixtures, and kitchen cabinets. This figure is a sketch plan of the house shown in isometric in Fig. 4-4. Compare the two figures to get an idea of shape, size, and relationship of the rooms. Try to imagine what it would be like to stand in the center of each room. Where would the light come from? Is there adequate ventilation? Is there privacy in the bedroom area? Is the living room, dining room, and kitchen well planned for family use?

34 Building Trades Blueprint Reading

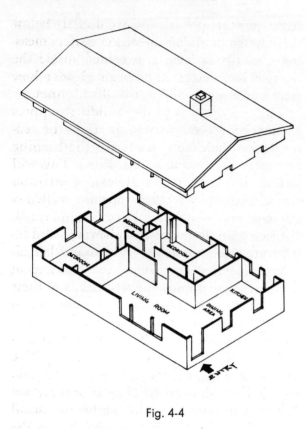

Fig. 4-4

A pictorial sketch of a small home cut by a horizontal plane. Fig. 4-3 shows the floor plan.

placed to provide light, ventilation, and to permit the arrangement of furniture. Doors have been omitted in both Figs. 4-3 and 4-4. Notice, however, that if they had been drawn in place they would have swung back against walls so as to take a minimum amount of space from each room. Closets and storage space are very important in making a house livable. A coat closet which will also serve as a windbreak and room divider is suggested near the entry. This is shown with hidden lines because it would not be a built-in feature. The closets for Bedrooms 1 and 2 are arranged so that they fit side by side. The closet for Bedroom 3 could be a wardrobe with sliding or bi-fold doors. The closet near the bathroom could be used for linens, or could be used for the location of a heating plant.

The bathroom fixtures have all been arranged along one wall with the kitchen sink on the other side of the same wall. All of the plumbing is therefore enclosed in the same partition.

Two Story House. The house shown in Fig. 4-5 has two stories and a full basement. The three cutting planes which produce the basement, first floor, and second floor plans are indicated by lines designated *A-A, B-B* and *C-C* respectively. Notice that the slice designated *A-A*, which becomes the projection for the basement plan, passes through the basement windows. The other two planes pass through the house at the level of the upper sash of windows and the upper part of doors.

A set of blueprints should be studied in much the same order as a person might inspect the house. The first floor plan should be observed first. See Fig. 4-6. A perspective drawing, Fig. 4-7, shows the same plan view using a three-dimensional illusion. Use these two figures to learn about the layout. Windows have been oversimplified and doors have been omitted in order that attention may be directed toward major features.

A very quick view of the first floor plan shows that the living and dining rooms are actually one large L-shaped room and that the kitchen, lavatory, and rear exit take up the remaining quarter of the whole area. Stairs to the second floor begin from a platform one riser above the floor level. The stairway is open with a railing on the living room side. Stairs to the basement descend

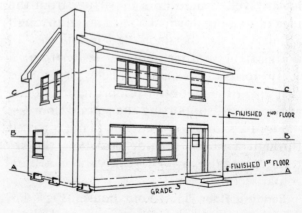

Fig. 4-5

A two story house with basement will have three floor plans. Notice the location of cutting planes A-A, B-B, and C-C.

Reading Floor Plans 35

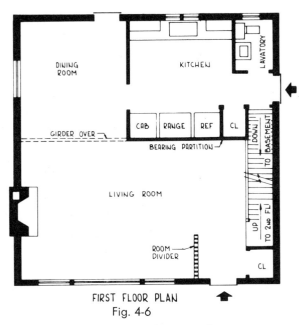

FIRST FLOOR PLAN
Fig. 4-6

The layout of partitions and location of openings, stairs, fireplace, fixtures, and cabinets is the function of a sketch floor plan. The cutting plane which passes through the building to produce the floor plan is taken at a level of 4 or 5 feet except at the fireplace. In order to show the fireplace construction the cutting plane is taken at a height of 1 foot above the floor. (This is the first floor plan of the house shown in Fig. 4-5.)

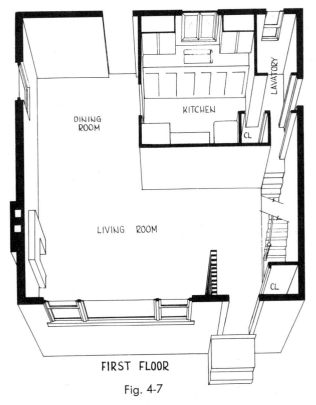

FIRST FLOOR
Fig. 4-7

A perspective view of the first floor helps in understanding the first floor plan shown in Fig. 4-6.

from the rear hallway and are directly below the stairs to the second floor. As a safety measure a small platform is provided inside the doorway before the stairs begin. Notice how both sets of stairs stop against break lines.

The construction of the building requires support for second floor joists across the center of the building. (Joists are the framing members which support the floor.) This will be done by the use of a load-bearing partition and a built-up structural member called a girder which will be placed as indicated. Kitchen equipment, such as the range and refrigerator, are shown on the floor plan although they are not part of the building in this particular house. A careful study of their location must be made by the architect so that they may be situated in the most convenient place and so that electrical outlets and gas connections are located exactly. The floor plan shows the location of the base and top kitchen cabinets. Additional information about cabinets is usually shown in detail views. The fireplace is drawn as though the plane passed about one foot above the floor so that details of its shape will show on the drawing. Fig. 4-6. The flue shown is for the basement furnace. (In Fig. 4-7, for illustrative purposes, the fireplace is drawn as if the plane passed through the wall above the shelf at the top of the fireplace. Thus both the furnace flue and the fireplace flue is shown. Fig. 4-6 shows the actual practice.)

The second floor plan, Fig. 4-8, and its accompanying perspective view, Fig. 4-9, show the layout of second floor rooms. The top riser of the stairway is shown in its exact location, and treads descend until they stop against a break line. The stairway is open to the hall and protected by a railing. Light is provided for the hall and stairs by a window at the head of the stairs. The bathroom is located above the lavatory on the first floor to provide economy in piping. The bathroom fixtures are shown by simplified symbols. Closets are shown with sliding doors so that no space is taken by swinging or folding doors. A load-bearing partition, which is directly over the girder shown on the first floor

SECOND FLOOR PLAN

Fig. 4-8

Second floor partitions are arranged to provide maximum use of space. (This is the second floor plan of the house shown in Fig. 4-5.)

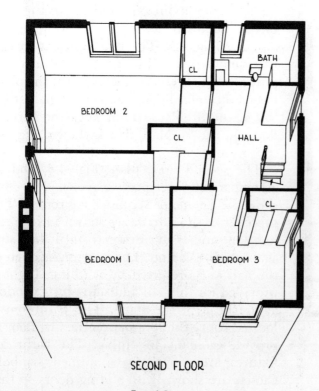

SECOND FLOOR

Fig. 4-9

A perspective view shows the layout of the second floor. Fig. 4-8 shows the floor plan.

plan, and other necessary structural members (not shown) which continue to the wall over the stairs, support the joists overhead. The chimney shows two flues, one for the fireplace on the first floor and the other for the furnace in the basement. Before it passes the 2nd floor level, the chimney is contained within the wall and projects on the outside only. Bedrooms 1, 2, and 3 are provided with cross ventilation.

The basement plan, Fig. 4-10, and the accompanying perspective view, Fig. 4-11, show some information about the foundation, windows, stairs, and the building structure. The location of the bottom riser is shown so that the carpenter may build the stairs to the first floor correctly. Again some of the treads are drawn but terminate against a break line. A steel beam is directly below the load-bearing partition on the first floor and supports the joists which run at right angles to it. The steel beam does not continue to the wall at the stairs because it would pass through the stairwell.

The fireplace foundation is carried down to the footing (support base of foundation wall) with only the furnace flue showing. A sufficient number of windows must be provided to give adequate ventilation and light. Areaways permit them to be partially below grade. Note also the symbol for showing 1st floor joists and the notation about their size on the basement floor plan, Fig. 4-10.

One-and-One-Half Story House. A one floor house with a roof of sufficiently high slope so that usable attic space is provided may be called a one-and-one-half story house. The traditional Cape Cod house would fall into this class. Dormers are generally used to add to the usable floor area, to provide ventilation, and to add to the architectural effect. They are usually described by the type of roof used. Fig. 4-12 shows several types and Fig. 4-13 shows a house with dormers on the front and the rear of the roof. For structural reasons, whenever it is possible to do so, the face wall of the dormer should be placed directly over the wall of the first floor.

A modified Cape Cod house, Fig. 4-13, is

Reading Floor Plans 37

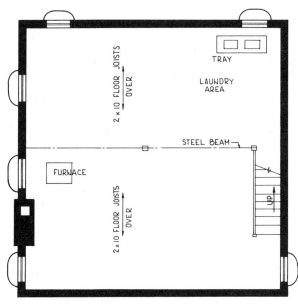

BASEMENT FLOOR PLAN

Fig. 4-10

The basement plan gives information on the foundation and structure of the house and other details. (This is the basement plan of the house shown in **Fig. 4-5**.)

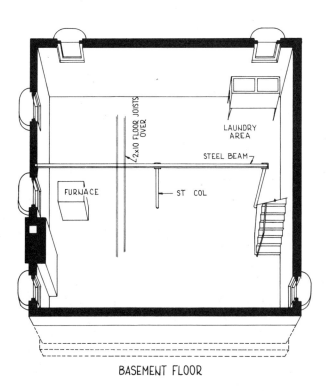

BASEMENT FLOOR

Fig. 4-11

Details of the basement are easily understood when presented in perspective form. **Fig. 4-10** shows the basement plan.

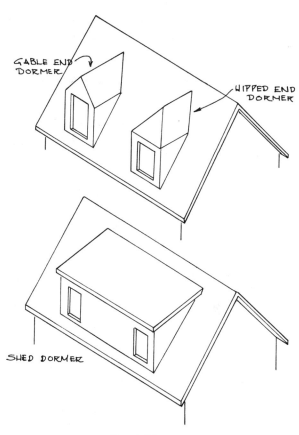

Fig. 4-12

Dormers of various types provide usable space, ventilation, and light.

Fig. 4-13

A modified Cape Cod house uses attic space for living.

designed with two gable end dormers on the front and a shed dormer which extends almost the full width on the rear of the house. The second floor plan, Fig. 4-14, is a slice taken by passing a plane through walls and partitions between four to five feet above the floor. Notice how the roof is shown wherever it appears at the sides and in front of the dormers. A hidden (dashed) line indicates that the face of the building on the first floor and that of the dormers is identical. The roof extends beyond the hidden (dashed) line on the front and rear of the building to form the eaves, and projects beyond the face of the wall on the gable ends of the house. The small triangular-shaped spaces left where the main roof and the floor meet may be used for storage when they are large enough to warrant such use.

The perspective view of the house, Fig. 4-15, may be useful in studying the details. Use it with the floor plan, Fig. 4-14, to gain a complete understanding of the layout. The stairs ascend to the second floor at the center of the building. The stairway is closed on both sides by partitions. Natural lighting is provided by a window at the head of the stairs. The bedrooms have adequate light and ventilation provided by windows in the gable ends of the house and in the dormers. Bedroom 2 has a built-in wardrobe instead of the conventional closet. The massive chimney projects slightly into this room. Rooms such as these can have a great deal of charm, even though the floor plan is irregular and part of the ceiling follows the slope of the roof.

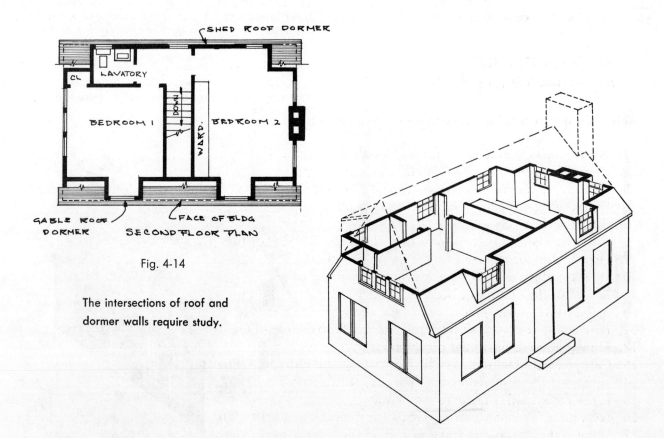

Fig. 4-14

The intersections of roof and dormer walls require study.

Fig. 4-15

A one-and-one-half story house is cut by a plane 4 to 5 feet above the floor to produce the floor plan.

Reading Floor Plans 39

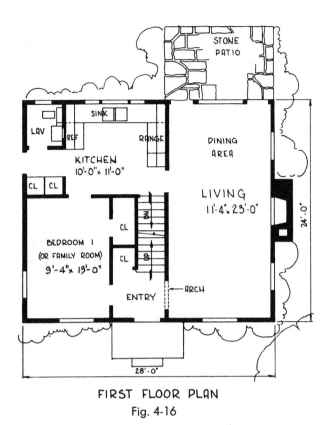

FIRST FLOOR PLAN
Fig. 4-16

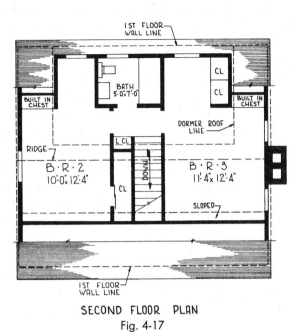

SECOND FLOOR PLAN
Fig. 4-17

This is a one-and-one-half story house with several features of the Cape Cod house in Figs. 4-13 and 4-14. These are not working drawings. A new idea has been introduced in showing the size of the major rooms. Closets, bathrooms, hallways and entry are not counted as rooms.

Self-Check Quiz No. 4

(Based on Figs. 4-16 and 4-17)

Answer the questions briefly. *Answers are given in the appendix.*

1. Why is this considered a one-and-a-half story house? _____
2. Does this house have a basement? _____
3. How many rooms does this house have not including lavatory, bath and closets? _____
4. Does the front of the house have a dormer? _____
5. Does the rear of the house have a dormer? _____
6. Is the face of the rear dormer directly above the face of the first floor wall line? _____
7. How is the dormer roof indicated on the plan? _____
8. Does the dormer roof have a projection at the eaves? _____
9. Is part of the ceiling in bedrooms 2 and 3 sloped? _____
10. Is the front entrance doorway in the center of the front elevation? _____
11. How many risers are shown from the grade to the top of the front entrance platform? _____
12. How is an arch indicated on the plan? _____
13. Can you tell how many risers there are on the stairs to the basement? _____
14. What are the dimensions of the living-dining area? _____
15. Is the patio made of concrete or stone? _____
16. How many exterior doorways are shown on the first floor plan? _____
17. What is the purpose of the two flues in the second floor chimney? _____
18. How many closets are shown on the second floor? _____
19. Which bedroom is the largest? _____
20. Is the space under the roof at the rear corners of the second floor accessible for storage? _____

TEST DRAWING NO. 3 (See pages 41-42)

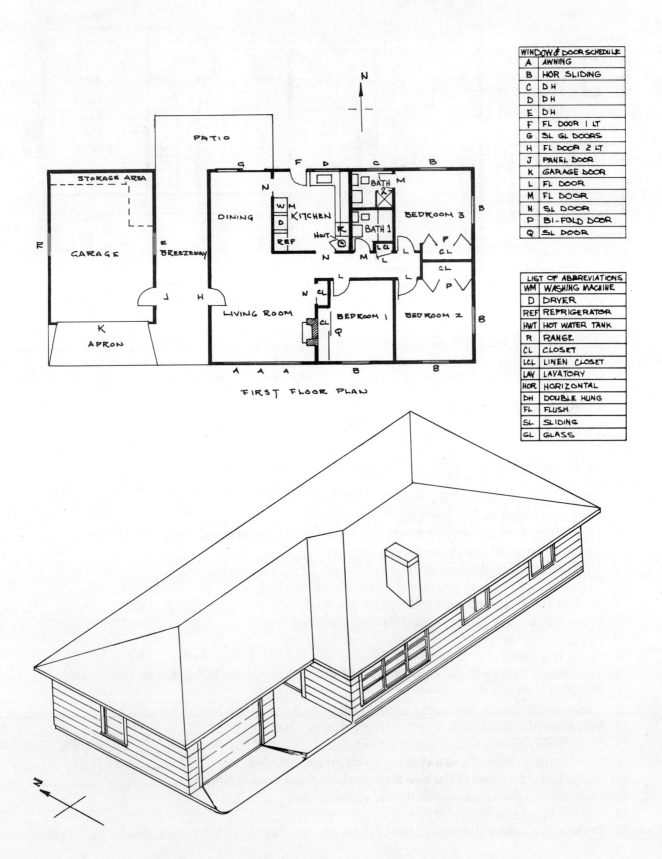

Reading Floor Plans 41

Trade Competency Test No. 3
(Chapter 4: Based on Test Drawing No. 3)

STUDENT'S NAME

INSTRUCTOR'S NAME

MULTIPLE CHOICE TEST: The following statements are incomplete or are in the form of a question to be answered. Choose the item which will best complete the statement or answer the question and place the letter in the space provided.

Example:

	Answer	Score
The kitchen door will be shown on which elevation? A. North.　　C. East. B. South.　　D. West.	A	
1. What type of roof does the house have? 　A. hip roof.　　C. shed roof. 　B. gable roof.　　D. shed and hip roof.		
2. The house has 　A. a full basement.　　C. attic storage space. 　B. no basement.　　D. two levels.		
3. The living room 　A. is separated from the dining room by a room divider. 　B. has direct access to the kitchen. 　C. has a fireplace on the west wall. 　D. is part of a living-dining room area.		
4. The dining area 　A. has direct access to the breezeway. 　B. has a swinging door to the kitchen. 　C. has glass sliding doors to the patio. 　D. has casement windows.		
5. The kitchen 　A. has cabinets on three sides. 　B. has two doors. 　C. has an exit to the patio. 　D. serves also as a laundry.		
6. The hallway 　A. opens to all of the rooms but bath #2 and dining room. 　B. has no door which swings outward into it. 　C. is L-shaped. 　D. has all swinging doors.		
7. Regarding closets, 　A. bedroom no. 1 has two closets. 　B. the linen closet has a sliding door. 　C. bedrooms no. 2 and 3 have closets of equal size. 　D. no closet is provided in the kitchen.		

Tear Off Here

(Cont.)

Building Trades Blueprint Reading

Trade Competency Test No. 3 *(Cont.)*

8. Regarding plumbing,
    A. all of the plumbing fixtures in the bathroom are arranged along the same wall.
    B. bath #2 has a tub.
    C. the kitchen sink is located under the kitchen window.
    D. the washing machine is located on the east kitchen wall.

9. The south elevation will have what type of windows?
    A. double hung and fixed sash.
    B. awning type and horizontal sliding.
    C. hopper and casement.
    D. double hung and casement.

10. The east elevation will have what type of windows?
    A. double hung.   C. awning.
    B. casement.   D. horizontal sliding.

11. The north elevation will have what type of windows?
    A. horizontal sliding and double hung.
    B. horizontal sliding and casement.
    C. double hung and casement.
    D. awning and double hung.

12. Regarding exterior doors for the house F, G and H,
    A. they are all alike.
    B. two are flush doors and one pair sliding doors.
    C. F and G are flush doors with the same light arrangement.
    D. F and G have locks on the same sides.

13. Regarding interior doors,
    A. all are swinging or sliding.
    B. all closet doors are sliding.
    C. all closet doors are swinging, sliding, or bi-fold.
    D. all doors to rooms from the hall are swinging.

14. Regarding the breezeway and patio,
    A. the breezeway and garage are under the same roof.
    B. the breezeway may be seen from the dining room.
    C. the patio and breezeway have separate concrete slabs.
    D. the patio extends past the kitchen door.

15. The garage
    A. floor and apron are level.
    B. has a gable roof.
    C. has two doors and two windows.
    D. has no structural tie with the house.

Tear Off Here

SCORE:

Chapter 5

Symbols and Notations Used on Floor Plans

It is essential that the architect and the men who do the work on the job understand each other. On the one hand it is the responsibility of the architect to give as much information on the drawings as needed. On the other hand the craftsman must know how to read everything on the drawings so that he understands what involves him directly and understands where his work and that of other men of the building team are inter-related. The architect has three problems to solve. (1) He is working at a small scale and must omit lines in order that the blueprints may be readable. Symbols and conventions must be used. (2) He does not have a great deal of space in which to letter information and thus uses abbreviations wherever he can. He is always alert to the fact that the builders must be able to understand them. (3) The time element in preparing the working drawings must be considered. It would take too much time and be repetitious to draw building parts in detail and to letter words and notations in full. Also this would not add to the information to be transmitted.

Symbols for Material and Building Parts. The symbols shown on Fig. 5-1 are used to represent the common materials used in buildings. Some of them resemble the thing they represent, but in most cases they must be learned without this relationship. Fig. 3-1, page 17, has shown how some of these materials are used "in elevation," such as on the exterior of the building. Fig. 5-1 shows how material is used on elevation views; on floor plan views; and on sectional views, such as a wall cut by a vertical plane or slice. Study these symbols and refer to them frequently in order to learn them. Fig. 5-2 shows variations and combinations of materials used on plan views. Fig. 5-3 shows how windows, doors, and other building parts are shown on working drawings. It is important to be able to recognize the symbols for the various types of windows and doors and also to know how several types of material may be used in the same wall.

Architects do their best to follow the accepted standards in representing materials, equipment, and building parts. Occasionally more than one symbol may be used. New materials are always being developed requiring new symbols and designations. The American National Standards Institute, (ANSI) and many other associations and trade groups, each interested in certain areas of the building business, are working constantly to standardize procedures. They are very much concerned with symbols. Fig. 5-4 on electrical symbols and Fig. 5-5 on plumbing symbols are examples of the work of these groups.

Abbreviations. The architect feels compelled to use abbreviations for the same reason he uses symbols to represent materials. He is limited in space and he must conserve time. There is a constant danger, however, that the person who reads the blueprints may not be able to understand some of the abbreviations because they are new or unfamiliar. The architect has the responsibility to use the accepted form, and whenever there may be a doubt about interpretation he should spell out the words. The craftsman should try his best to learn the terms in their shortened form and whenever he is puzzled by some new or strange term, he should find out what it means before proceeding with the job.

Some abbreviations have two or more meanings. "D" for instance can stand for "drain" or "dryer." "FL" can mean "flashing," "floor," or "flush." Usually there is a clue given by the fact that a leader and an arrowhead points to something specific on the blueprint. Self Check Test 5-B will bring out the idea that abbreviations are almost always associated with something tangible. The list of abbreviations contains most of the common abbreviations shown on floor plans. There are several more which might be included but which are used less frequently. (Common elevation symbols were given in Chapter 3.) Only capital letters are shown because most draftsmen use capital letters exclusively. Periods are usually omitted.

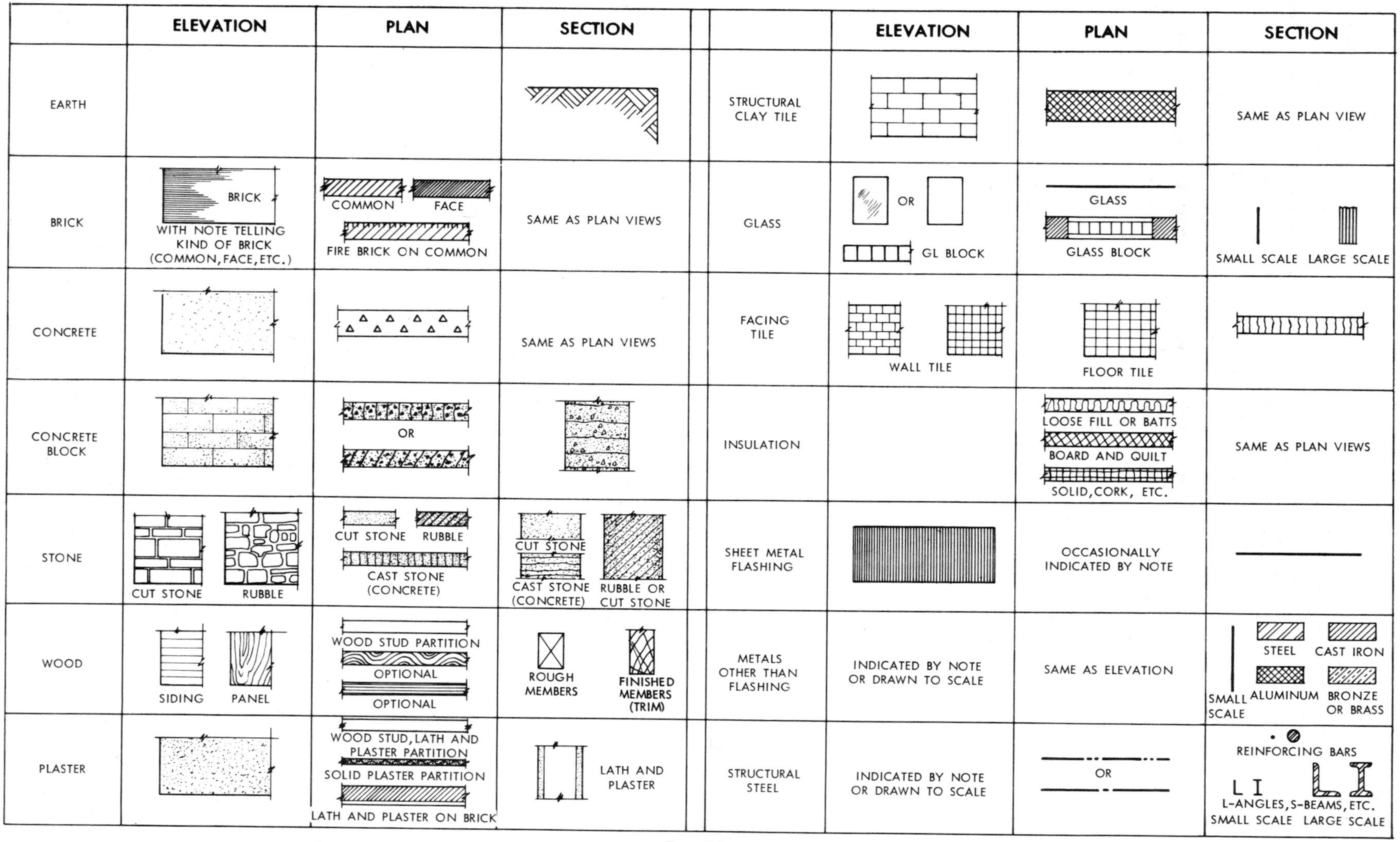

Symbols and Notations Used on Floor Plans **45**

ELEVATION	PLAN	SECTION		ELEVATION	PLAN	SECTION
EARTH						SAME AS PLAN VIEW
BRICK — WITH NOTE TELLING KIND OF BRICK (COMMON, FACE, ETC.)	COMMON / FACE / FIRE BRICK ON COMMON		STRUCTURAL CLAY TILE			SAME AS PLAN VIEW
CONCRETE		SAME AS PLAN VIEWS	GLASS	OR / GL BLOCK	GLASS / GLASS BLOCK	SMALL SCALE / LARGE SCALE
CONCRETE BLOCK	OR	SAME AS PLAN VIEWS	FACING TILE	WALL TILE	FLOOR TILE	
STONE — CUT STONE / RUBBLE	CUT STONE / RUBBLE / CAST STONE (CONCRETE)	CUT STONE / CAST STONE (CONCRETE) / RUBBLE OR CUT STONE / FINISHED MEMBERS (TRIM) / ROUGH MEMBERS	INSULATION		LOOSE FILL OR BATTS / BOARD AND QUILT / SOLID, CORK, ETC.	SAME AS PLAN VIEWS
WOOD — SIDING / PANEL	WOOD STUD PARTITION / OPTIONAL / OPTIONAL		SHEET METAL FLASHING	INDICATED BY NOTE OR DRAWN TO SCALE	OCCASIONALLY INDICATED BY NOTE	
PLASTER	WOOD STUD, LATH AND PLASTER PARTITION / SOLID PLASTER PARTITION / LATH AND PLASTER ON BRICK	LATH AND PLASTER	METALS OTHER THAN FLASHING	INDICATED BY NOTE OR DRAWN TO SCALE	SAME AS ELEVATION	
			STRUCTURAL STEEL	INDICATED BY NOTE OR DRAWN TO SCALE	OR	STEEL / CAST IRON / SMALL ALUMINUM BRONZE SCALE OR BRASS / REINFORCING BARS / L-ANGLES, S-BEAMS, ETC. SMALL SCALE LARGE SCALE

Fig. 5-1 Symbols used for common materials shown on blueprints.

46 Building Trades Blueprint Reading

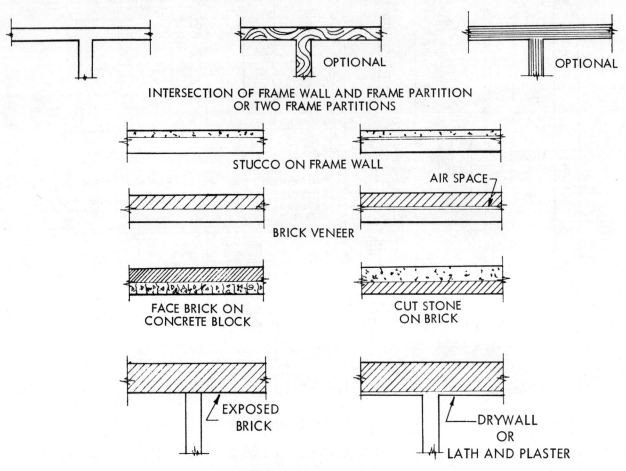

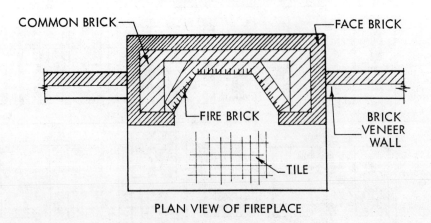

Fig. 5-2

Materials are often used in combination.

Symbols and Notations Used on Floor Plans 47

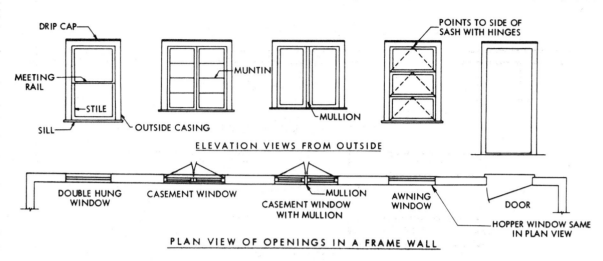

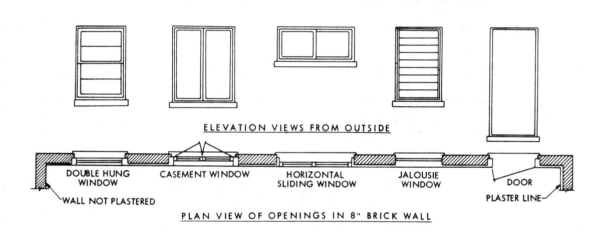

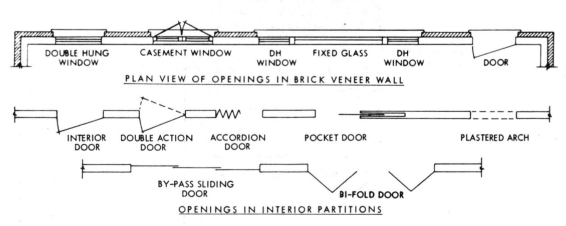

Fig. 5-3 Symbols commonly used for openings in exterior walls and interior partitions. The different types of wall construction will be discussed in Chapter 7.

48 Building Trades Blueprint Reading

Self-Check Quiz No. 5-A on Symbols for Material
(Based on Figs. 5-1 and 5-2)

Fill in the blanks with the correct names. *Answers are given in the appendix.*

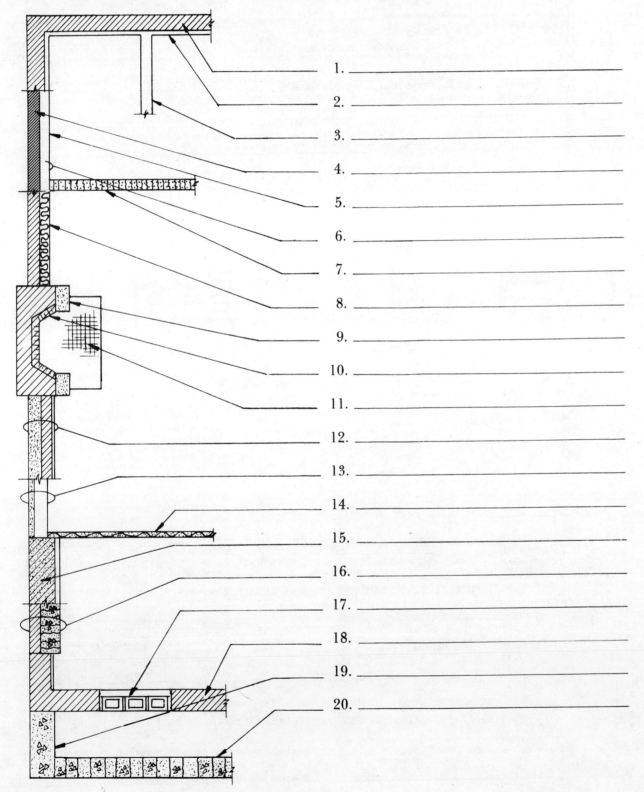

1. _____
2. _____
3. _____
4. _____
5. _____
6. _____
7. _____
8. _____
9. _____
10. _____
11. _____
12. _____
13. _____
14. _____
15. _____
16. _____
17. _____
18. _____
19. _____
20. _____

Symbols and Notations Used on Floor Plans 49

General Outlets

- Lighting Outlet
- Ceiling Lighting Outlet for recessed fixture (Outline shows shape of fixture.)
- Continuous Wireway for Fluorescent Lighting on ceiling, in coves, cornices, etc. (Extend rectangle to show length of installation.)
- Lighting Outlet with Lamp Holder
- Lighting Outlet with Lamp Holder and Pull Switch
- Fan Outlet
- Junction Box
- Drop-Cord Equipped Outlet
- Clock Outlet

To indicate wall installation of above outlets, place circle near wall and connect with line as shown for clock outlet.

Convenience Outlets

- Duplex Convenience Outlet
- Triplex Convenience Outlet (Substitute other numbers for other variations in number of plug positions.)
- Duplex Convenience Outlet — Split Wired
- Weatherproof Convenience Outlet
- Multi-Outlet Assembly (Extend arrows to limits of installation. Use appropriate symbol to indicate type of outlet. Also indicate spacing of outlets as X inches.)
- Combination Switch and Convenience Outlet
- Combination Radio and Convenience Outlet
- Floor Outlet
- Range Outlet
- Special-Purpose Outlet. Use subscript letters to indicate function. DW-Dishwasher, CD-Clothes Dryer, etc.
- Protected by Ground Fault Circuit Interrupter
- Smoke Detector

Switch Outlets

- S — Single-Pole Switch
- S_3 — Three-Way Switch
- S_4 — Four-Way Switch
- S_D — Automatic Door Switch
- S_P — Switch and Pilot Light
- S_{WP} — Weatherproof Switch
- S_2 — Double-Pole Switch

Low-Voltage and Remote-Control Switching Systems

- $\underline{S}$ — Switch for Low-Voltage Relay Systems
- $\underline{MS}$ — Master Switch for Low-Voltage Relay Systems
- O_R — Relay—Equipped Lighting Outlet
- — - — - — Low-Voltage Relay System Wiring

Auxiliary Systems

- Push Button
- Buzzer
- Bell
- Combination Bell-Buzzer
- CH — Chime
- Annunciator
- D — Electric Door Opener
- M — Maid's Signal Plug
- Interconnection Box
- T — Bell-Ringing Transformer
- Outside Telephone
- Interconnecting Telephone
- R — Radio Outlet
- TV — Television Outlet

Miscellaneous

- Service Panel
- Distribution Panel
- ---- Switch Leg Indication. Connects outlets with control points.
- $O_{a,b}$, etc. — Special Outlets. Any standard symbol given above may be used with the addition of subscript letters to designate some special variation of standard equipment for a particular architectural plan. When so used, the variation should be explained in the Key of Symbols and, if necessary, in the specifications.
- ——— Branch Circuit; Concealed in Ceiling or Wall.
- --- Branch Circuit; Concealed in Floor.
- ----- Branch Circuit; Exposed.
- Home Run to Panel Board. Indicate number of Circuits by number of arrows. Note: Any circuit without further designation indicates a two-wire circuit. For a greater number of wires indicate as follows —///— (3 wires) —////— (4 wires), etc.
- ——— Feeders. Note: Use heavy lines and designate by number corresponding to listing in Feeder Schedule.

American National Standards

Fig. 5-4   Electrical Symbols for Architectural Plans

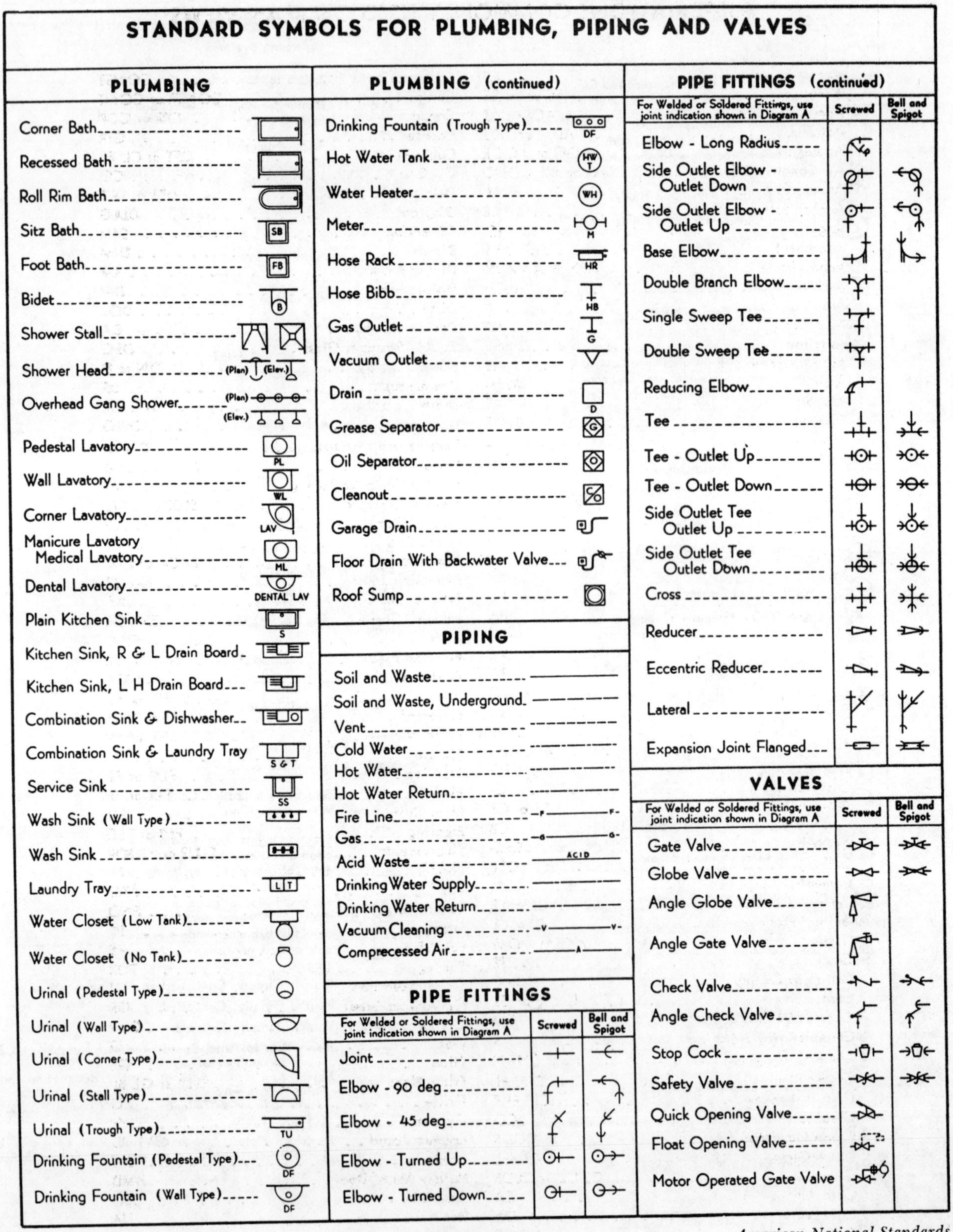

50 Building Trades Blueprint Reading

STANDARD SYMBOLS FOR PLUMBING, PIPING AND VALVES

PLUMBING

- Corner Bath
- Recessed Bath
- Roll Rim Bath
- Sitz Bath — SB
- Foot Bath — FB
- Bidet — B
- Shower Stall
- Shower Head (Plan) (Elev.)
- Overhead Gang Shower (Plan) (Elev.)
- Pedestal Lavatory — PL
- Wall Lavatory — WL
- Corner Lavatory — LAV
- Manicure Lavatory / Medical Lavatory — ML
- Dental Lavatory — DENTAL LAV
- Plain Kitchen Sink — S
- Kitchen Sink, R & L Drain Board
- Kitchen Sink, L H Drain Board
- Combination Sink & Dishwasher
- Combination Sink & Laundry Tray — S & T
- Service Sink — SS
- Wash Sink (Wall Type)
- Wash Sink
- Laundry Tray — LT
- Water Closet (Low Tank)
- Water Closet (No Tank)
- Urinal (Pedestal Type)
- Urinal (Wall Type)
- Urinal (Corner Type)
- Urinal (Stall Type)
- Urinal (Trough Type) — TU
- Drinking Fountain (Pedestal Type) — DF
- Drinking Fountain (Wall Type) — DF

PLUMBING (continued)

- Drinking Fountain (Trough Type) — DF
- Hot Water Tank — HWT
- Water Heater — WH
- Meter — M
- Hose Rack — HR
- Hose Bibb — HB
- Gas Outlet — G
- Vacuum Outlet
- Drain — D
- Grease Separator
- Oil Separator
- Cleanout
- Garage Drain
- Floor Drain With Backwater Valve
- Roof Sump

PIPING

- Soil and Waste
- Soil and Waste, Underground
- Vent
- Cold Water
- Hot Water
- Hot Water Return
- Fire Line ——F——F——
- Gas ——G——
- Acid Waste ——ACID——
- Drinking Water Supply
- Drinking Water Return
- Vacuum Cleaning ——V——V——
- Compreccessed Air ——A——

PIPE FITTINGS

For Welded or Soldered Fittings, use joint indication shown in Diagram A	Screwed	Bell and Spigot
Joint	+	←
Elbow - 90 deg		
Elbow - 45 deg		
Elbow - Turned Up		
Elbow - Turned Down		

PIPE FITTINGS (continued)

For Welded or Soldered Fittings, use joint indication shown in Diagram A	Screwed	Bell and Spigot
Elbow - Long Radius		
Side Outlet Elbow - Outlet Down		
Side Outlet Elbow - Outlet Up		
Base Elbow		
Double Branch Elbow		
Single Sweep Tee		
Double Sweep Tee		
Reducing Elbow		
Tee		
Tee - Outlet Up		
Tee - Outlet Down		
Side Outlet Tee Outlet Up		
Side Outlet Tee Outlet Down		
Cross		
Reducer		
Eccentric Reducer		
Lateral		
Expansion Joint Flanged		

VALVES

For Welded or Soldered Fittings, use joint indication shown in Diagram A	Screwed	Bell and Spigot
Gate Valve		
Globe Valve		
Angle Globe Valve		
Angle Gate Valve		
Check Valve		
Angle Check Valve		
Stop Cock		
Safety Valve		
Quick Opening Valve		
Float Opening Valve		
Motor Operated Gate Valve		

American National Standards

Fig. 5-5

Symbols used for plumbing fixtures, piping, fittings, and valves.

Symbols and Notations Used on Floor Plans

ABBREVIATIONS COMMONLY USED ON PLAN VIEWS

Access Panel	AP
Acoustic	AC or ACST
Acoustical Tile	ACT or AT
Adjustable	ADJT or ADJ
Aggregate	AGG or AGGR
Air Conditioning	A/C or AIR COND
Aluminum	AL
Anchor Bolt	AB
Angle	∠
Apartment	APT
Approximate	APX or APPROX
Architectural	ARCH
Area	A
Area Drain	AD
Asbestos	ASB
Asbestos Board	AB
Asphalt	ASPH
Asphalt Tile	AT
Basement	BSMT
Bathroom	B
Bath Tub	BT
Beam	BM
Bearing Plate	BPL or BRG PL
Bedroom	BR
Blocking	BLKG
Blueprint	BP
Boiler	BLR
Book Shelves	BK SH
Brass	BRS
Brick	BRK
Bronze	BRZ
Broom Closet	BC
Building	BLDG
Building Line	BL
Cabinet	CAB.
Calking	CK or CLKG
Casing	CSG
Cast Iron	CI
Cast Stone	CST or CS
Catch Basin	CB
Ceiling	CLG
Cellar	CEL
Cement	CEM
Cement Asbestos Board	CEM AB
Cement Floor	CEM FL
Cement Mortar	CEM MORT
Center	CTR
Center to Center	C to C
Center Line	℄ or CL
Center Matched	CM
Ceramic	CER
Channel	CHAN
Cinder Block	CIN BL
Circuit Breaker	CIR BKR
Cleanout	CO
Clean Out Door	CODR
Clear Glass	CL GL
Closet	C, CL or CLOS
Cold Air	CA
Cold Water	CW
Collar Beam	COL B
Concrete	CONC
Concrete Block	CONC BLK
Concrete Floor	CONC FLR or CONC FL
Concrete Masonry Unit	CMU
Conduit	CND
Construction	CONST
Contract	CONTR or CONT
Copper	CPR or COP
Counter	CTR
Cubic Feet	CFT or CU FT
Cut Out	CO
Detail	DTL or DET
Diagram	DIAG
Dimension	DIM
Dimmer	DIM
Dining Room	DR
Dishwasher	DW
Ditto	DO.
Double-Acting	DA
Double Strength Glass	DSG
Down	DN or D
Downspout	DS
Drain	D or DR
Drawing	DWG
Dressed and Matched	D & M
Dryer	D
Electric Metallic Tubing	EMT
Electric Operator	ELECT. OPR.
Electric Panel	EP
End to End	E to E
Excavate	EXCA or EXC
Expansion Joint	EXP JT
Exterior	EXT
Exterior Grade	EXT GR
Finish	FIN.
Finished Floor	FIN FLR or FIN. FL
Firebrick	FBRK
Fireplace	FPL or FP
Fireproof	FP or FPRF
Fixed Window	FX WDW
Fixture	FIX.
Flashing	FLG or FL
Floor	FLR or FL
Floor Drain	FD
Flooring	FLR or FLG
Fluorescent	FLUR or FLUOR
Flush	FL
Footing	FTG
Foundation	FND
Frame	FR
Full Size	FS
Furring	FUR
Galvanized Iron	GI
Galvanized Steel	GS
Garage	GAR
Gas	G
Glass	GL
Glass Block	GLB or GL BL
Grille	G
Gypsum	GYP
Gypsum Board	GYP BD
Hardware	HDW
Hollow Metal Door	HMD
Hose Bibb	HB
Hot Air	HA
Hot Water	HW
Hot Water Heater	HWH
Inside Diameter	ID
Insulation	INS
Interior	INT
Iron	I

ABBREVIATIONS COMMONLY USED ON PLAN VIEWS (Cont'd)

Jamb	JB
Kitchen	KIT or K
Landing	LDG
Lath	LTH
Laundry	LAU
Laundry Tray	LT
Lavatory	LAV
Leader	L
Length	L, LG or LGTH
Library	LIB
Light	LT
Limestone	LMS or LS
Linen Closet	L CL
Lining	LN
Linoleum	LINO
Living Room	LR
Louver	LVR or LV
Main	MN
Marble	MRB or MR
Masonry Opening	MO
Material	MTL or MATL
Maximum	MAX
Medicine Cabinet	MC
Minimum	MIN
Miscellaneous	MISC
Mixture	MIX
Modular	MOD
Mortar	MOR
Moulding	MLD or MLDG
Nosing	NOS
Obscure Glass	OBSC GL
On Center	OC
Open Web Joist	OJ or OW JOIST
Opening	OPG or OPNG
Outlet	OUT
Overall	OA
Overhead	OH or OVHD
Pantry	PAN.
Partition	PTN
Per Square Inch	PSI
Plaster	PLAS or PL
Plastered Opening	PO
Plate	PL
Plate Glass	PG or PL GL
Platform	PLAT
Plumbing	PLBG
Porch	P
Precast	PRCST
Prefabricated	PFB or PREFAB
Pull Switch	PS
Quarry Tile	QT
Radiator	RAD
Random	RDM
Range	R
Recessed	REC
Refrigerator	REF
Register	REG
Reinforce or Reinforcing	RE or REINF
Reinforcing Steel Bar	RE BAR
Revision	REV
Riser	R
Roof	RF
Roof Drain	RD
Room	RM or R
Rough	RGH
Rough Opening	RO or RGH OPNG
Rubber Tile	RBT or R TILE
S Beam	S
Scale	SC
Schedule	SCH
Screen	SCN or SCR
Scuttle	S
Section	SEC or SECT
Select	SEL
Service	SERV
Sewer	SEW.
Sheathing	SHTH or SHTHG
Sheet	SHT or SH
Shelf and Rod	SH & RD
Shelving	SH or SHELV
Shower	SH
Sill Cock	SC
Single Strength Glass	SSG
Sink	SK or S
Sliding Door	SL DR
Soil Pipe	SP
Specification	SPEC
Square Feet	SQ FT
Stained	STN
Stairs	ST
Stairway	STWY
Standard	STD
Steel	ST or STL
Steel Sash	SS
Storage	STO or STG
Structural Clay Products Research Foundation	SCR
Switch	SW or S
Telephone	TEL
Tempered Plate Glass	TEM PL GL
Terra Cotta	TC
Terrazzo	TZ or TER
Thermostat	THERMO
Threshold	TH
Toilet	T
Tongue and Groove	T & G
Tread	TR or T
Typical	TYP
Unfinished	UNF
Unexcavated	UNEXC
Utility Room	U RM
Vent	V
Vent Stack	VS
Vestibule	VEST
Vinyl Tile	VT or V TILE
Vitreous Tile	VIT TILE
Wainscot	WSCT or WAIN.
Warm Air	WA
Washing Machine	WM
Water	W
Water Closet	WC
Water Heater	WH
Water Resistant	WR
Waterproof	WP
Weather Stripping	WS
Weephole	WH
Welded Wire Fabric	WWF
White Pine	WP
Wide Flange	W or WF
Wood	WD
Wood Frame	WF
Yellow Pine	YP

Symbols and Notations Used on Floor Plans   53

Self-Check Quiz No. 5-B Plan Abbreviations

(Based on abbreviations, pages 51-52)

Study the list of abbreviations shown on pages 51-52. In the space provided fill in the full word for the abbreviated word without referring to the list. *Answers are given in the appendix.*

Example: BSMT STAIRWAY   BASEMENT

1. AL SASH   1. _____
2. 16" C TO C   2. _____
3. OAK FIN. FLR   3. _____
4. FIBER GLASS INS   4. _____
5. REC RADIATOR   5. _____
6. WALL HUNG W C   6. _____
7. DRAIN TO C B   7. _____
8. C W SUPPLY   8. _____
9. SERVICE RACEWAY TO E P   9. _____
10. 3'-6" M O FOR FRAME   10. _____
11. SH & RD IN CLOSET   11. _____
12. UNEXC UNDER KITCHEN   12. _____
13. 4" CEM FLR   13. _____
14. CONC B WALL   14. _____
15. EXCA FOR FOOTING   15. _____
16. P S LIGHT   16. _____
17. WINDOW WITH SSG   17. _____
18. EQUAL FROM CTR   18. _____
19. ½" BITUMINOUS EXP JT   19. _____
20. RECESSED MC   20. _____
21. S P TO SEWER   21. _____
22. 4 × 10 BM   22. _____
23. CER TILE WALL   23. _____
24. INT TRIM   24. _____
25. OBSC GL IN DOOR   25. _____

26. REF NICHE   26. _____
27. ½" BPL   27. _____
28. F S DET   28. _____
29. REG FACE   29. _____
30. 3 WAY SW   30. _____
31. CIN BL PARTITION   31. _____
32. D A DOOR   32. _____
33. 14 R UP   33. _____
34. CHIMNEY CO DR   34. _____
35. DSG IN SIDE LIGHTS   35. _____
36. GI FLASHING   36. _____
37. LAV FAUCETS   37. _____
38. PLAS ARCH   38. _____
39. RBT IN ENTRY   39. _____
40. BRZ TH AT FRONT DOOR   40. _____
41. C A RETURN   41. _____
42. H B 1'-6" FROM GRADE   42. _____
43. 2 × 4 PL   43. _____
44. W GIRDER   44. _____
45. SLOPE TO DR   45. _____
46. CABINET HDW   46. _____
47. LINO FLOOR   47. _____
48. FINISHED P O   48. _____
49. T & G SIDING   49. _____
50. W S WINDOWS   50. _____

54 Building Trades Blueprint Reading

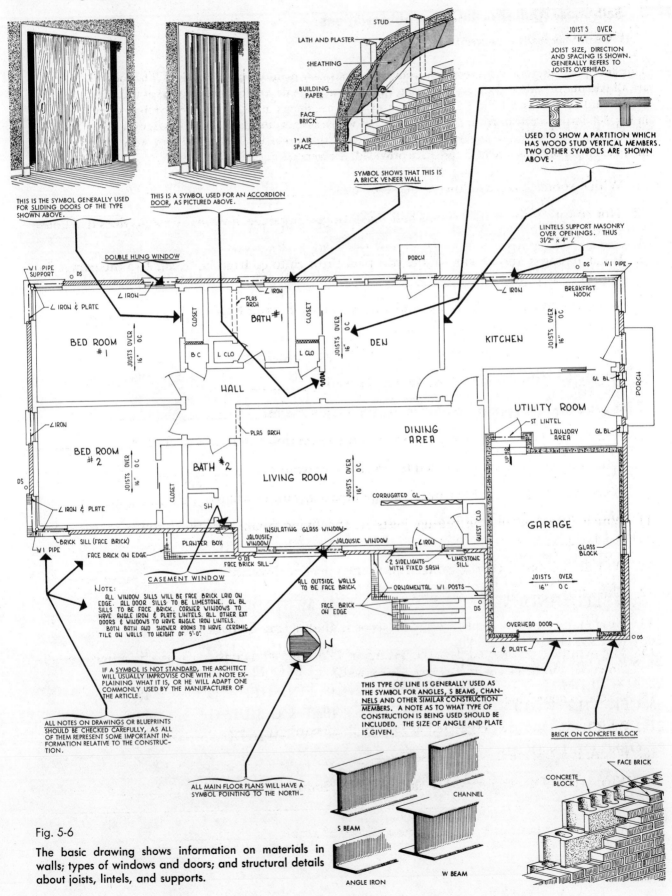

Fig. 5-6

The basic drawing shows information on materials in walls; types of windows and doors; and structural details about joists, lintels, and supports.

Symbols and Notations Used on Floor Plans 55

Self-Check Quiz No. 5-C
(Based on Fig. 5-6)

This floor plan is several steps removed from the preliminary plans used in Chapter 4. The rooms are all laid out to scale and the partitions and outside walls are drawn to show the type of materials used according to Figs. 5-1 and 5-2. Windows and doors are shown in place using the symbols given in Fig. 5-3. Structural features are added. The joists which support the floor above, lintels over windows and doors, and wrought iron pipe supports at the corner windows are shown. Answer the following questions briefly in the space provided. *Answers are given in the appendix.*

1. Which rooms have windows which face east? _____

2. Not counting the utility room, hall, bathrooms, or garage, how many rooms does this house have? _____

3. What passages and rooms would you pass through to go from bedroom 1 to the garage by the shortest route? _____

4. Is the whole floor of the house and garage the same level? Explain _____
_____

5. What type of exterior walls are shown for the building? _____

6. What type walls are shown for the garage walls? _____

7. What type of partitions are shown between main rooms? _____

8. What type of partition is shown between the linen closet and the den closet? _____

9. What kinds of members are used for lintels over windows? _____

10. What is used to support the lintels at the bedroom corner windows? _____

11. Which direction do the ceiling joists in the living room run, north-south or east-west? _____

12. Why do the joists over the garage run in the opposite direction? _____

13. What is the spacing of the joists? _____

14. How many exterior doors are shown (not including garage door)? _____

15. The window designated Double Hung in bedroom #1 is typical. Is the window symbol shown in the brick or frame part of the wall? _____

16. Where is glass block used? _____

17. Which windows are designated as having fixed sash? _____

18. Where is a casement window used? _____

19. What types of doors are used in the house (excluding the garage)? _____
_____

20. What kind of lines are used to show a plastered arch? _____

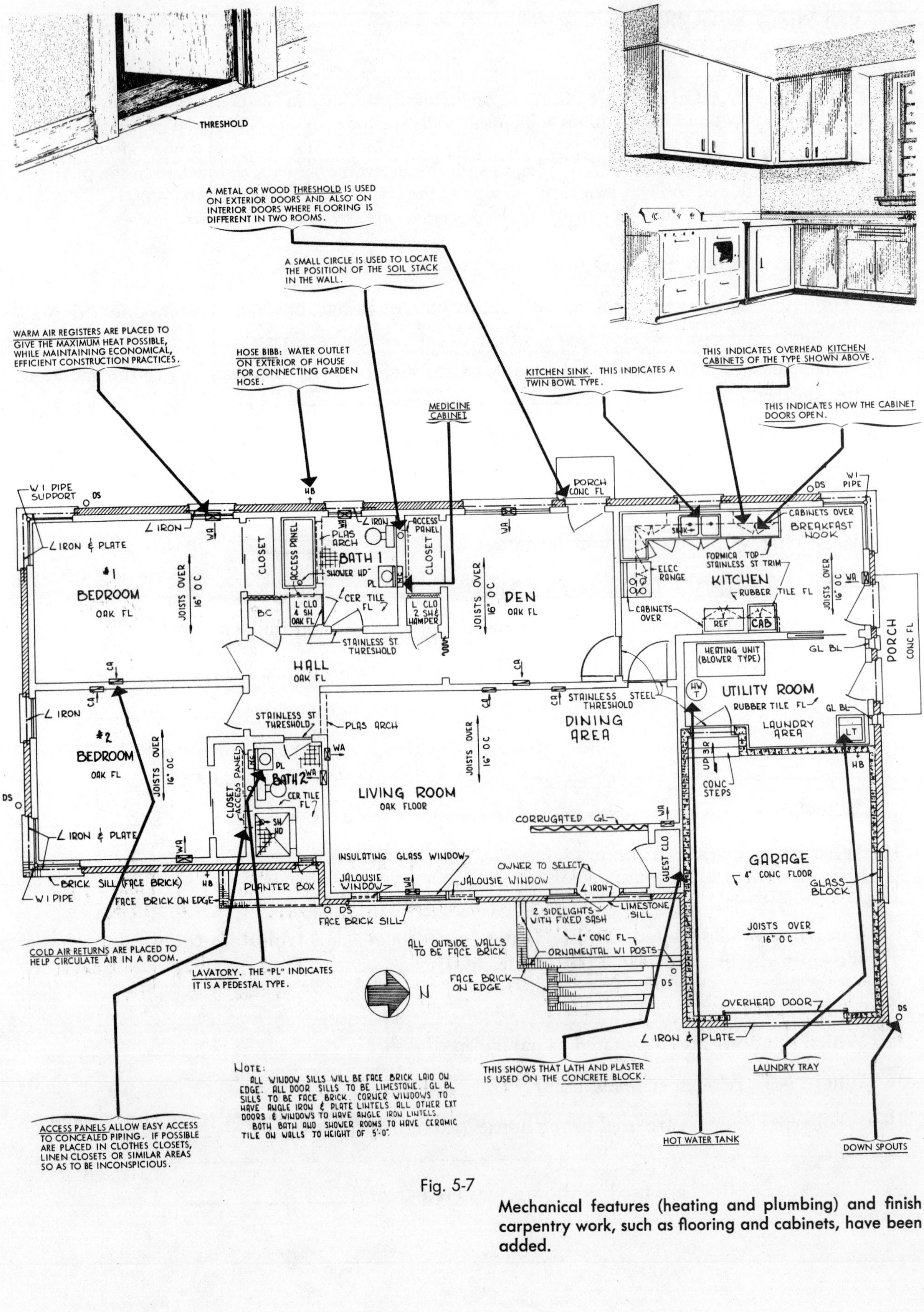

56 Building Trades Blueprint Reading

Fig. 5-7

Mechanical features (heating and plumbing) and finish carpentry work, such as flooring and cabinets, have been added.

Symbols and Notations Used on Floor Plans

Self-Check Quiz No. 5-D
(Based on Fig. 5-7)

The information added to the basic floor plan in Fig. 5-7 includes all of the plumbing fixtures and the heating plant showing the location of warm air and cold air registers. Information about the finish is included, such as flooring, cabinets, and shelving. Answer the following questions in the spaces provided. *Answers are given in the appendix.*

1. Where is the hot water tank located? _____

2. Soil stacks are located in partitions behind which fixtures? _____

3. How many hose bibbs are shown? _____

4. What does LT mean? _____

5. What are access panels for? _____

6. Where are medicine chests located in the bathrooms? _____

7. Is the heating plant hot water or forced warm air? _____

8. How many warm air registers are there? _____ How many cold air registers? _____

9. Which rooms do not have cold air returns? _____

10. Why are most of the warm air registers placed on outside walls? _____
_____

11. Why is no register indicated in the utility room? _____

12. Is there any heat provided in the garage? _____

13. Is there a built-in range or oven? _____

14. Are cabinets indicated over the refrigerator? _____

15. How many stainless steel thresholds are shown? _____ What are they for? _____
_____

16. What trim is built in to the linen closet nearest the den? _____

17. What is the corrugated glass partition in the living room-dining area for? _____

18. What kind of flooring is used in the kitchen and utility room? _____

19. What kind of flooring is used in the bathrooms and shower? _____

20. What kind of flooring is used in the rest of the house? _____

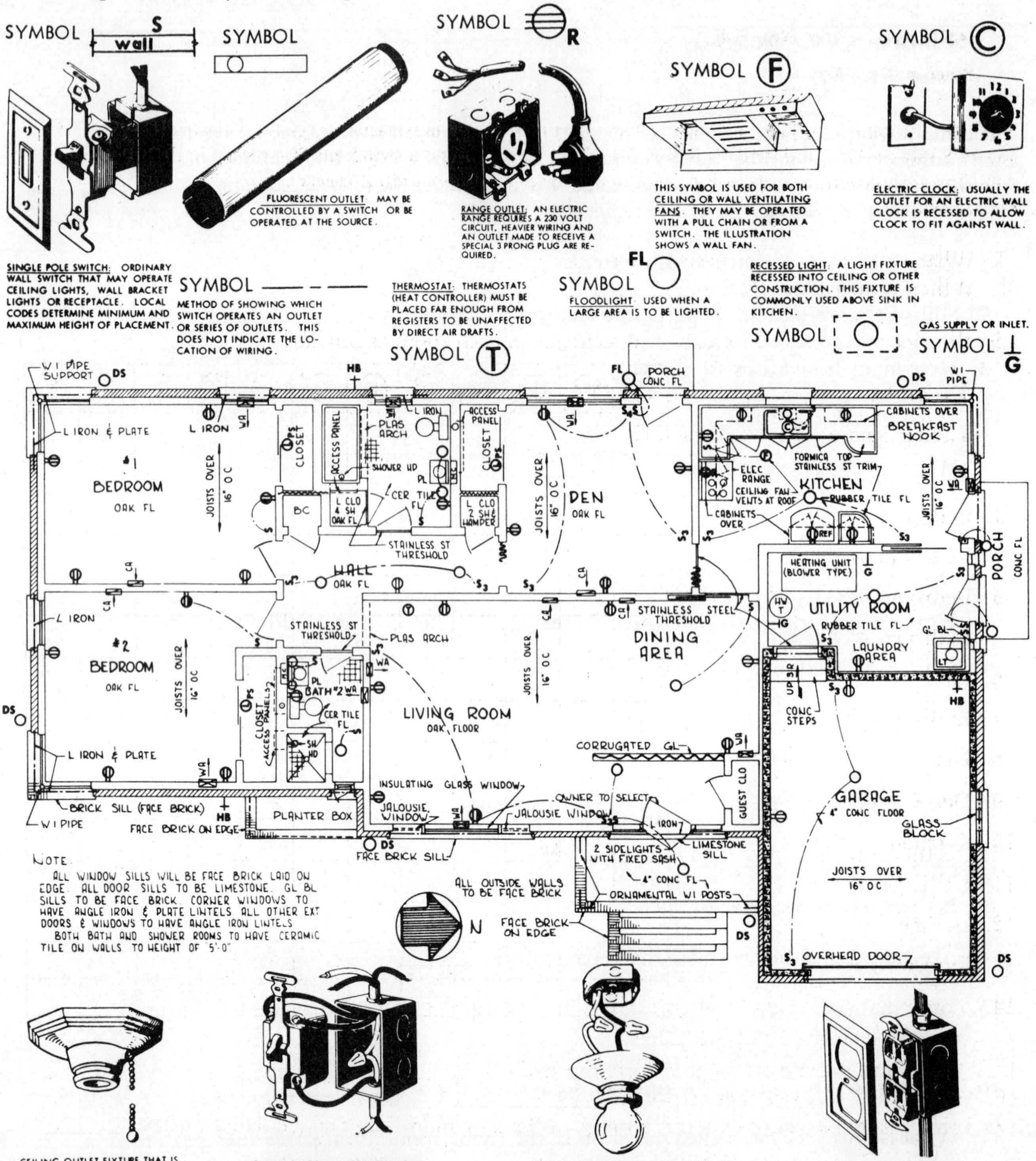

58 Building Trades Blueprint Reading

SYMBOL S wall

SINGLE POLE SWITCH: ORDINARY WALL SWITCH THAT MAY OPERATE CEILING LIGHTS, WALL BRACKET LIGHTS OR RECEPTACLE. LOCAL CODES DETERMINE MINIMUM AND MAXIMUM HEIGHT OF PLACEMENT.

SYMBOL

FLUORESCENT OUTLET: MAY BE CONTROLLED BY A SWITCH OR BE OPERATED AT THE SOURCE.

SYMBOL ⊖ R

RANGE OUTLET: AN ELECTRIC RANGE REQUIRES A 230 VOLT CIRCUIT. HEAVIER WIRING AND AN OUTLET MADE TO RECEIVE A SPECIAL 3 PRONG PLUG ARE REQUIRED.

SYMBOL F

THIS SYMBOL IS USED FOR BOTH CEILING OR WALL VENTILATING FANS. THEY MAY BE OPERATED WITH A PULL CHAIN OR FROM A SWITCH. THE ILLUSTRATION SHOWS A WALL FAN.

SYMBOL C

ELECTRIC CLOCK: USUALLY THE OUTLET FOR AN ELECTRIC WALL CLOCK IS RECESSED TO ALLOW CLOCK TO FIT AGAINST WALL.

SYMBOL - - -

METHOD OF SHOWING WHICH SWITCH OPERATES AN OUTLET OR SERIES OF OUTLETS. THIS DOES NOT INDICATE THE LOCATION OF WIRING.

THERMOSTAT: THERMOSTATS (HEAT CONTROLLER) MUST BE PLACED FAR ENOUGH FROM REGISTERS TO BE UNAFFECTED BY DIRECT AIR DRAFTS.

SYMBOL T

SYMBOL FL

FLOODLIGHT: USED WHEN A LARGE AREA IS TO BE LIGHTED.

RECESSED LIGHT: A LIGHT FIXTURE RECESSED INTO CEILING OR OTHER CONSTRUCTION. THIS FIXTURE IS COMMONLY USED ABOVE SINK IN KITCHEN.

SYMBOL ▢

GAS SUPPLY OR INLET.

SYMBOL ⊥ G

SYMBOL L PS

CEILING OUTLET FIXTURE THAT IS CONTROLLED BY A PULL CHAIN.

3-WAY SWITCH: USED WHEN ONE OR MORE OUTLETS ARE CONTROLLED AT TWO OR MORE PLACES. IF THE OUTLET OR OUTLETS ARE CONTROLLED AT THREE PLACES, ONE 4-WAY SWITCH IS USED WITH THE OTHER TWO 3-WAY SWITCHES.

SYMBOL S₃

CEILING OUTLET: THIS SYMBOL SHOWS THE FIXTURE IS CONTROLLED BY A SWITCH OR SWITCHES.

SYMBOL ○

A THREE PRONG GROUNDED OUTLET. ALL RECEPTACLES INSTALLED ON 15 AMPERE AND 20 AMPERE CIRCUITS SHALL BE OF A GROUNDED TYPE.

SYMBOL ⊖

Fig. 5-8

Information on gas and electricity has been added to the floor plan. Only dimensions are lacking to make this a working drawing.

Symbols and Notations Used on Floor Plans 59

Self-Check Quiz No. 5-E
(Based on Fig. 5-8)

Fig. 5-8 is the same floor plan as shown in Figs. 5-6 and 5-7 with information on gas and electricity added. Study the electrical symbols in Fig. 5-4. Note: an outlet and a switch may be placed in the same box as shown in the garage, Fig. 5-8. Count each as a separate unit. *Answers are given in the appendix.*

1. What is the symbol for gas supply? _____
2. Which units require gas? _____

List the electrical outlets, fixtures and switches in each room as indicated below.

Example:	⊙	⊖	ⓛ$_{PS}$	⊖$_R$	⊖	Ⓕ	Ⓒ	⊙$_{FL}$	Ⓣ	S	S$_3$	S$_4$
Living Room—Dining Area	2	6		1					1	2	2	
3. Bathroom #2												
4. Bedroom #1												
Bedroom #1 Closet												
5. Bedroom #2												
Bedroom #2 Closet												
6. Hall												
7. Bathroom #1												
8. Den												
9. Den Closet												
10. Kitchen												
11. Utility Room												
12. Garage												
13. Exterior Lights												
14. Total												

NOTE: Include example in total.

15. Why is the thermostat located near the hall? _____
16. When are 3-way switches required? _____
17. What is the difference between a range outlet and an ordinary convenience outlet? _____
_____
18. Is the electric panel (fuses and switches) shown? _____
19. What do these designations mean when placed with outlet or switch symbols: PS, FL? _____
_____
20. When outlets are controlled at three places, what type of switch is used with two 3-way switches? _____

TEST DRAWING NO. 4 (See pages 61-62)

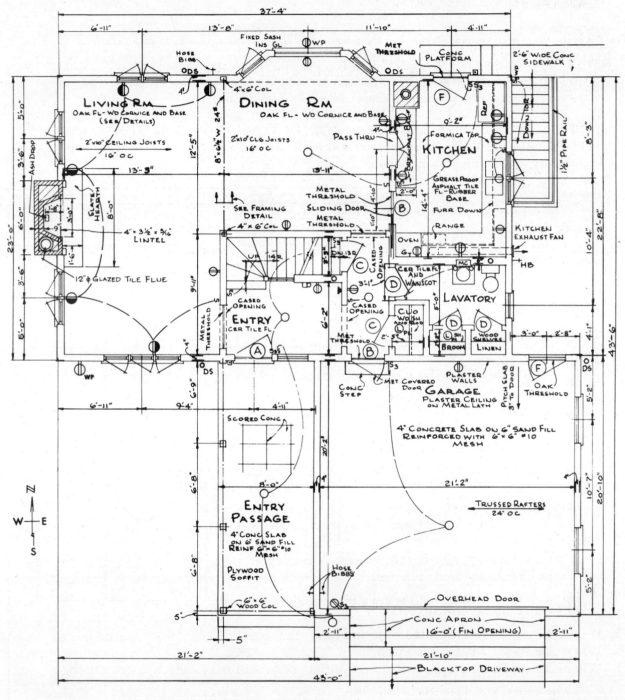

FIRST FLOOR PLAN SCALE ¼"=1'-0"

NOTE: ALL DIMENSIONS ARE TO FACE OF STUDS

PUBLISHER'S NOTE: THIS DRAWING WAS ORIGINALLY DRAWN TO THE SCALE SHOWN. THE DRAWING WAS REDUCED TO FIT THE PAGE AND CAN NO LONGER BE SCALED.

Door Schedule

Mark	Size	Am't Req'd	Remarks	Mark	Size	Am't Req'd	Remarks
A	3'-0"x6'-8"x1¾"	1	Exterior Flush Door	D	2'-4"x6'-8"x1⅜"	4	Flush Doors
				D₁	2'-4"x6'-8"x1⅜"	1	Louvered
B	2'-8"x6'-8"x1¾"	7	Flush Doors 1-Sliding 1-Metal Covered	E	1'-3"x6'-8"x1⅜"	1	Bi-Fold Louvered
C	2'-6"x6'-8"x1⅜"	4	Flush Doors	F	2'-10"x6'-8"x1¾"	2	Exterior 2-Lights
C₁	2'-6"x6'-8"x1⅜"	2	Louvered	G	2'-8"x6'-8"x1¾"	1	Exterior 2-Lights

Trade Competency Test No. 4
(Chapter 5: Based on Test Drawing No. 4)

STUDENT'S NAME

INSTRUCTOR'S NAME

NOTE: Refer to the simplified floor plan of the house in order to become thoroughly familiar with the layout of the rooms (Fig. 4-1, p. 31). Also refer to the blueprint elevation drawings Figs. 3-7, 3-8, and 3-9 to gain a clear idea of what the house will look like when built.

MULTIPLE CHOICE TEST: The following statements are incomplete. After each statement you will find four phrases or sentences. Choose the one which will best complete the sentence and place the corresponding letter in the answer column.

	Answer	Score
Example: The exterior walls of the building are of A. frame construction. C. stucco on frame construction. B. brick veneer construction. D. brick construction.	A	

1. This first floor plan shows
  A. a complete bathroom.
  B. an enclosed entry passage.
  C. an open living room-dining room combination.
  D. a separate garage.
	1.	

2. Access to the second floor is gained by
  A. a stairway partially open to the entry.
  B. a stairway open to the living room.
  C. a stair with 13 risers.
  D. a stair which makes a 180° turn.
	2.	

3. Access to the basement is gained by
  A. winding stair from the entry.   C. exterior stairway.
  B. stair with 14 risers.   D. stairway open to entry.
	3.	

4. Regarding floor levels,
  A. the living and garage areas are the same level.
  B. the garage floor is lower than the living area.
  C. the garage and entry passage floor levels are the same.
  D. the garage floor is level.
	4.	

5. The windows in the living room and dining room are all
  A. casement windows.
  B. casement windows with mullions.
  C. insulating glass.
  D. casement or fixed sash.
	5.	

6. The kitchen and lavatory windows
  A. are double hung.
  B. are casement.
  C. are casement with mullion, or double hung.
  D. are casement without a mullion or double hung.
	6.	

(Cont.)

Trade Competency Test No. 4 *(Cont.)*

7. Doors designated B on this plan are
   A. swinging or sliding (pocket).
   B. panel doors.
   C. of different sizes.
   D. metal clad.

8. Doors designated C and D are
   A. louvered.
   B. identical.
   C. all 1¾″ thick.
   D. all alike except for size and direction of swing.

9. Metal thresholds are provided
   A. only at outside doors.
   B. only at doors where flooring changes from one material to another.
   C. at lavatory doorway.
   D. at garage doors.

10. Regarding flooring,
    A. the house has either oak or ceramic tile.
    B. the garage floor and the entry passage floor have the same material and base.
    C. the fireplace hearth is tile.
    D. the entry has a slate floor.

11. Regarding the living room-dining room area,
    A. this area is provided with a wood base and plaster cornice.
    B. this area has no ceiling outlets.
    C. this area has all convenience outlets controlled by wall switches.
    D. a pass-through is provided to the kitchen.

12. Closets
    A. have louvered doors.
    B. have shelves and rods.
    C. have shelves.
    D. have lights with pull switches.

13. The kitchen
    A. has a counter top range.
    B. has ceramic tile floor and base.
    C. has an electric oven.
    D. has upper cabinets across two walls.

14. The lavatory
    A. has one ceiling light.
    B. has ceramic tile floor and base only.
    C. has a soil stack located in the partition behind the toilet.
    D. has one light at the medicine chest.

15. The following electrical outlets and switches have been omitted:
    A. push buttons and door bell.
    B. telephone.
    C. waterproof exterior outlet and switch.
    D. clock outlet.

SCORE:

Chapter 6

Scaling and Dimensional Practices

Drawing to Scale. The blueprints used on the job are reproductions of architect's plans and working drawings *drawn to scale*. The idea of drawing objects to scale and reading drawings which are drawn to scale is not new to anyone. Little children draw representations of houses, trees, and animals which are many times larger than the sheets of paper they have to use. As they attempt to draw them in relationship to each other, they are trying to apply a scale to their work.

A road map is a common example of a drawing made to scale. An area of several thousand square miles is shown on a piece of paper only a few square feet in area. It is drawn at a scale of a certain number of miles per inch.

If it were possible to make working drawings of a house at full size and then make blueprints from them, they would be of little use on the construction job because they would be extremely cumbersome. The drawings, therefore, are made small enough so that the blueprints can be handled easily and also large enough so that the blueprints show the necessary information clearly. The length of every line is reduced to a constant fraction of its true length so that all of the parts of the building are in exact relationship to each other. The scale which is usually used for floor plans and elevation drawings is $1/4'' = 1'-0''$. Wherever detailed drawings are required to show the arrangement of doors and drawers of kitchen cabinets, information on the construction of a front entrance doorway, etc., a scale of $1\frac{1}{2}'' = 1'-0''$ is often used.

A house which is 32 feet long and 24 feet wide would be drawn as a rectangle 8 inches by 6 inches if $1/4'' = 1'-0''$ scale were used:

$$32 \times 1/4'' = 8'' \qquad 24 \times 1/4'' = 6''$$

Since each *one-fourth* inch represents 1 foot, 32 feet would be represented by 32 one-fourth inches or 8 inches; and 24 feet would be represented by 24 one-fourth inches or 6 inches. In order to avoid calculating each dimension, the architect uses an *architect's scale* which gives him direct proportional readings in several scales.

The Architect's Scale. The architect's scale used most often in school is triangular with six ruled faces designed to measure in ten different scales. See Fig. 6-1. One of the edges is identical to a twelve inch ruler divided into sixteenth's of an inch. (Another type of architect's scale is flat with beveled edges which have only four ruled faces. There is no difference in the scales or in their application. With the triangular scale, however, there are more scales to choose from.)

One of the scales used in Fig. 6-1 is designated $1/4$ which means that it is intended to be used in making drawings at a scale of $1/4'' = 1'-0''$. One foot at this scale would be drawn an actual length of $1/4$ inch. (Note that $1/4$ inch is equal to $1/48$ of a foot; therefore, a drawing made with this scale is $1/48$ actual size.)

64 Building Trades Blueprint Reading

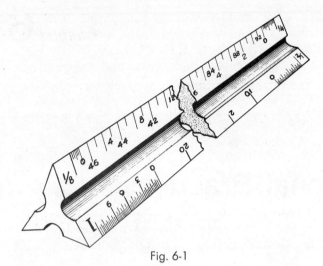

Fig. 6-1

An Architect's Scale: The scale is one foot long plus a small portion provided on each end so that the first units may be read clearly and accurately. The broken line near the center indicates that the center section of the drawing is omitted so that the figure does not take up too much space. Four scales are shown on this side. They are ⅛" = 1'-0", ¼" = 1'-0", ½" = 1'-0" and 1" = 1'-0".

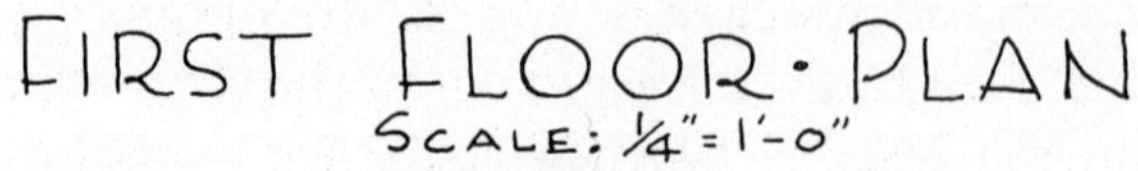

The scale to which a drawing is made is always indicated on the drawing. This scale is generally placed below the title, as shown in Fig. 6-2.

Following is a list of the various scales on an architect's scale:

Scale Labeled	Scale	Scale Labeled	Scale
3	3" = 1'	⅜	⅜" = 1'
1½	1½" = 1'	¼	¼" = 1'
1	1" = 1'	3/16	3/16" = 1'
¾	¾" = 1'	⅛	⅛" = 1'
½	½" = 1'	3/32	3/32" = 1'
		16	Full Scale

Although each of the above scales will be

FIRST FLOOR·PLAN
SCALE: ¼"= 1'-0"

Fig. 6-2

The scale to which a drawing is made is usually noted below the title of the drawing.

found on most architect's scales, their location may vary on the scales produced by different manufacturers.

Reading a Scale. Fig. 6-3 shows four scales. The ¼" = 1'-0" scale is read from right to left beginning with the line marked "0" near the right hand end. From this line, 46 spaces would represent a distance of 46 feet to scale. The ⅛" = 1'-0" scale is read from left to right beginning with a line marked "0" near the left hand end. From this line 92 spaces to the right would represent a distance of 92 feet at this scale. Thus the same set of markings is used for both the ⅛" = 1'-0" scale and the ¼" = 1'-0" scale. It is very important to read the numbers underneath the respective lines in relation to the scale used. Those related to the ⅛" = 1'-0" scale are in a row nearest the edge of the scale. Those related to the ¼" = 1'-0" scale are in the row farther from the edge of the scale. The lines for the units on this scale have been extended so that this distinction can be made.

In order to read measurements which are less than one foot on the ¼" = 1'-0" scale, Fig. 6-3, it is necessary to use the space to the right of the line marked "0." This space is

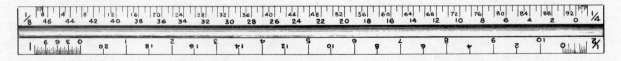

Fig. 6-3

This drawing shows one of the three sides of an Architect's Scale. Notice that the scale labeled "¼" reads from right to left (0 to 46) while the "⅛" scale reads from left to right (0 to 92).

divided into twelve parts, each representing one inch. Fig. 6-4 shows how the scale is used to read the dimension of the width of opening A-B. Full feet are measured on the left of the line marked "0" and the inches are measured on the right side of the line.

When a drawing is made at a scale of ⅛" = 1'-0" the scale marked ⅛ is used. Fig. 6-3. Whole feet are read to the right of the line marked "0" and parts of a foot are read to the left of the line. The space representing parts of a foot is divided into only six parts because, if it were divided into twelve parts, the spaces would be so small that they would be difficult to distinguish. Each one of the six parts represents 2 inches to scale.

Fig. 6-5 shows four examples of how distances are measured to scale using four different scales. The distances from A to B are to be determined in each case. The space used for measuring parts of a foot is divided into twelve parts on both the ¼" scale and the 3⁄16" scale. Each space represents one inch. The ½" scale and the ¾" scale are different in that the space is divided into 12 parts but each one is again divided in half. The smallest unit then represents one-half inch. Thus it is possible to measure spaces representing inches and also half inches.

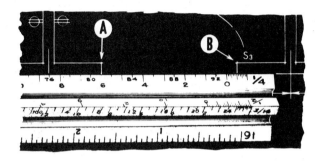

Fig. 6-4

To scale the distance from A to B, place the scale so that the largest possible number of feet on the scale is at Point A (6 in this case). There are 12 subdivisions on the scale to the right of 0; each one represents one inch. There are 6 subdivisions between 0 and Point B. The opening, therefore, is 6'-6".

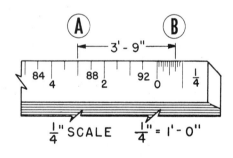

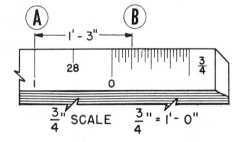

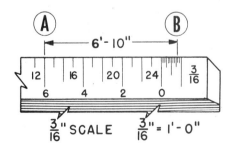

Fig. 6-5

The scale is placed so that full feet are measured on one side of the line marked "0" and inches on the other side of the line.

Notice that the 3⁄16" scale and ¼" scale provide divisions in feet and inches. The ½" scale and ¾" scale provide divisions in feet, inches, and half inches.

Self-Check Quiz No. 6-A Using an Architect's Scale
(Based on pages 63-65)

Using an architect's scale, measure the following line lengths using the scale given at the left. Insert the dimensions in the space provided. *Note* however that any measured or scaled dimensions on a print can be only approximate. Paper shrinkage, caused by the varying moisture content of the paper, can result in a line measurement that is off as much as two inches at $\frac{1}{4}'' = 1'\text{-}0''$. This is why it is a poor practice to *scale* a drawing. Always use the stated dimensions. *Answers are given in the appendix.*

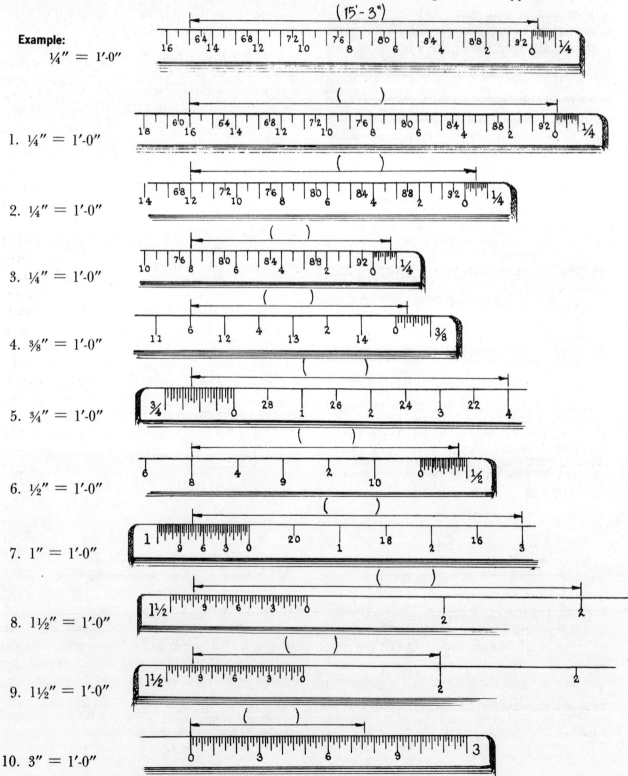

Scaling and Dimensioning Practices 67

Using Folding Rule to Draw to Scale and to Read Dimensions. A folding rule or any ruler divided into inches and 1/16's of an inch may be used to make drawings and to read drawings at 1/4" = 1'-0" scale. It must be stressed however that the results will not be perfectly accurate because the small divisions shown on the architect's scale are not available. Also, blueprints should be measured *only as a last resort* after every other means of obtaining the required dimension has been exhausted.

Occasionally a craftsman finds it necessary to make a sketch on the job to scale. This may be done in order to work out a problem, or to transmit information back to his shop or to the architect. The ability to use a folding rule as a device to make meaningful sketches will prove to be valuable.

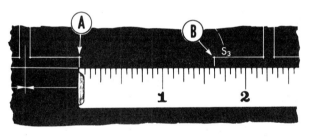

Fig. 6-6

A blueprint drawn at the scale of 1/4" = 1'-0" may be scaled using a folding rule. In measuring the distance from A to B, the end of the rule is placed on line A. The number of full 1/4" spaces is counted. There are six, each representing 1'-0". Two 1/16" spaces finish out the distance to point B. Each one represents 3 inches. Thus the dimension equals 6'-6".

Each 1/4 inch space on the folding rule is considered to be 1 foot on the sketch or blueprint. Each 1/16 inch space on the folding rule is considered to be 3 inches. Thus a distance of 1 3/16" would represent 4'-9" at the scale of 1/4" = 1'-0". It is necessary to count the number of 1/4 inch spaces to find the number of feet and add the number of 1/16" spaces giving each one the value of 3 inches. Fig. 6-6 gives an example of how a folding rule may be used to measure using the scale 1/4" = 1'-0".

Measuring Blueprints to Obtain Dimensions. Even though it is possible to measure blueprints with an architect's scale or folding rule to obtain missing dimensions, this practice should be used only if every other means has been used first. Blueprints are reproductions of original drawings and in the reproduction process a slight amount of shrinkage may occur. With the new diazo processes, this is negligible. However, a problem may develop in that the part of the drawing to be measured may not be drawn to scale. Architects do everything they can to draw the plans exactly to scale. Occasionally last minute changes are made which would mean completely redrawing the plans in order to have them exactly to scale. If the work must go ahead without taking this additional time, only the dimensions may be changed rather than doing all of the drafting work over again. When one dimension is changed usually it affects one or more other dimensions.

All the necessary dimensions are generally included on the blueprints. Sometimes it takes study to find them. They may be somewhere else on the same view or on one of the other blueprints in the set. Occasionally it is necessary to do a little arithmetic to find the information.

The over-all length of the house would only be shown on one of the sides. Heights of similar windows above the floor would be shown once on an elevation view. A series of dimensions locating partitions may be shown on one line across a floor plan. If partitions continue across the building from wall to wall, it is unnecessary to show the same dimensions in other rooms.

It may be necessary to look at several blueprints in order to find a desired dimension. A sectional view through an exterior wall, for example, may show dimensions which are required when working on the exterior of the house and which might otherwise appear on elevation views.

Self-Check Quiz No. 6-B Using a Folding Rule

(Based on page 67)

Using a folding rule or a ruler divided into 1/16ths, measure the distance between points. Insert the approximate dimensions in the space provided. Remember that paper shrinkage will cause line lengths to vary. SCALE: ¼" = 1'-0". *Answers are given in the appendix.*

1.
2.
3.
4.
5.
6.
7.
8.
9.
10.

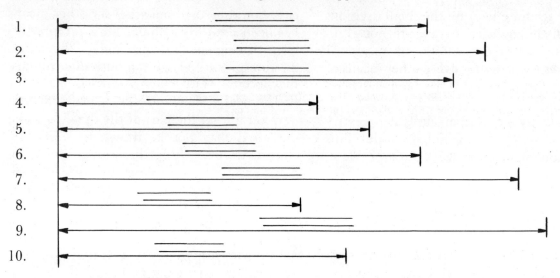

Self-Check Quiz No. 6-C Calculating and Checking Missing Dimensions

(Based on pages 67, 69)

Calculate the missing dimensions then scale them using a folding rule or ruler divided into 16ths. Insert the correct dimensions in the space provided. SCALE: ¼" = 1'-0". *Answers are found in the appendix.*

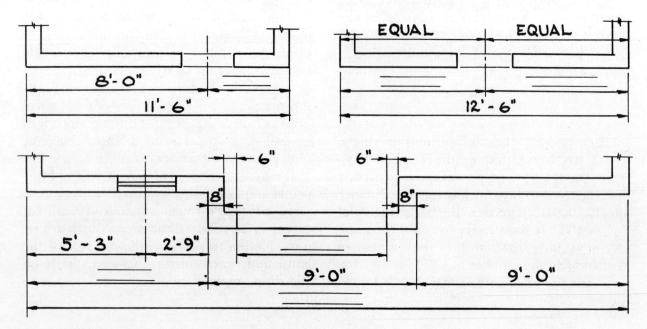

Scaling and Dimensioning Practices

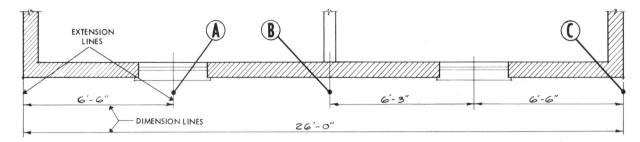

Fig. 6-7

Extension lines and dimension lines are thin, solid lines which are so placed that they cannot be misinterpreted as a part of the structure shown in the drawing. The arrows indicate the extreme ends of the distance for which a dimension is shown. Note that dimension lines are arranged so as not to cross each other.

The measurement from line A to line B may be obtained by calculation.

Some dimensions are obtained by simple arithmetic. Fig. 6-7 is an example of how to find a missing dimension when all but one dimension in a series are given. By adding the three dimensions given and subtracting them from the overall dimension, the distance from *A* to *B* can be obtained.

```
  6'- 6"               26'-0" =  25'-12"
  6'- 3"                        -19'- 3"
  6'- 6"                         -------
 -------                          6'- 9"
18'-15" or 19'-3"
```

The measurement from *A* to *B* should be 6'-9". (A review of arithmetic is included in the appendix.)

Drawing Symbols and Conventions to Scale. Every symbol or convention used on elevation or plan views is, as nearly as possible, drawn accurately to scale. In other words, walls, windows, doors, plumbing and other fixtures, footings, partitions, chimneys, roofs, etc., are all drawn in proportion to their size, following whichever of the various scales is being employed. Thus, if a wall is to be drawn 6 inches thick on a drawing which is drawn at ¼" = 1'-0" scale, the architect will draw the parallel lines of the wall symbol 6 inches apart to scale. Using a folding rule to scale the blueprint of the drawing, it would measure ⅛" (at ¼" = 1'-0" scale, ⅛" = 6 inches).

Bathroom fixtures are drawn to scaled sizes and located so that they fit into the space provided for them. Windows, doors, and all of the parts of the building are drawn to scale both on the plan and elevation views.

Fig. 6-8 is a plan view of the corner of a foundation and indicates the location of the footing with a hidden (dashed) line. Measuring the figure with an architect's scale will show that the foundation is 1'-0" wide and the footing is 2'-0" wide. If a folding rule is used instead, the foundation will be found to be ¼" wide or 1'-0" to scale, and the footing will be found to be ½" wide or 2'-0" to scale.

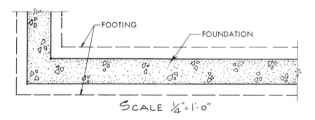

Fig. 6-8

Foundation with its footing, drawn to the ¼" scale.

Fig. 6-9 shows a door opening drawn to a scale of ⅜" = 1'-0". The wall is to be 6 inches thick and the opening is to be 3 feet. Check the figure with an architect's scale. (A folding rule does not lend itself very well to scales other than ¼" = 1'-0" for general application.)

70 Building Trades Blueprint Reading

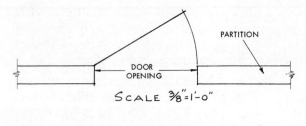

Fig. 6-9

Door in a frame wall drawn to the ⅜" scale.

Dimensioning Standards. Although there are projects continuously underway to standardize dimensioning procedures in the construction field, there are still many variations between different areas of the country, as well as between architects and draftsmen in the same area. Some of these variations would cause considerable trouble if they were overlooked in reading a print.

Exterior Walls. Fig. 6-10 (*A* and *B*) shows two ways of dimensioning a frame wall. The method shown at *A* is the recommended method, but is not used by all architects. The advantage of this system of dimensioning is that once the stud corner posts are located, most of the other parts of the structure can be located in terms of their distance from the corner of these rough framing members. For example, the doors and windows are framed before the sheathing is put on. (The boards or material applied to the rough frame work is called sheathing.) It is easier to locate the openings if dimensions start from the outside face of the studs at the corners than if an allowance must be made for the sheathing which has not been installed yet. If dimensions are shown as in Fig. 6-10*B*, the carpenter must subtract the thickness of the sheathing whenever he takes a measurement from the rough framework corner to locate a window or door opening. In this case, a note or designation stating that the dimensions are taken from the "outside face of sheathing" should appear on the blueprint so that this point is made clear.

Fig. 6-10*C* shows the standard dimensioning of solid masonry walls such as brick, tile, concrete block, stone, and poured concrete. Fig. 6-10*D* illustrates the standard dimensioning practice used with masonry veneer walls. Here, both the dimensions to the outside face of the studs and the dimensions to the outside of the masonry wall are given. The outside dimensions of the foundation usually are the same as the outside dimensions of the masonry part of the wall. The men who erect the forms for the foundation walls must have this information. It is also important that the

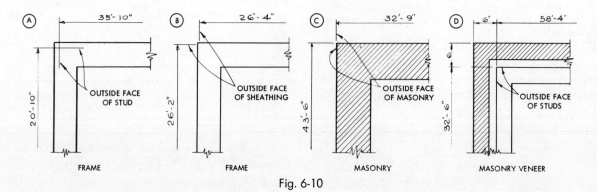

Fig. 6-10

"A" and "B" show two ways of dimensioning frame exterior walls at corners. Dimensions to the outside face of studs are more helpful to the builder than are dimensions to the outside face of sheathing. "C" shows how dimensions are taken from the corners of masonry buildings. "D" shows good practice in dimensioning masonry veneer structures. Both the frame part of the wall and the masonry part are fixed by measurements.

face of studs be located in relationship to the face of the masonry wall because the frame structure is erected before the masonry part.

Interior Partitions. Stud partitions vary in thickness depending on the finish. See Fig. 6-11. Dimensions are drawn to the center line as shown at E, Fig. 6-11, or to the face of the studs as shown at F. Wood stud partitions are considered to be 4" thick nominally. This does not include the wall finish whether it be lath and plaster, gypsum wall board (drywall), wood paneling, or other material. Because of building variations which develop on the job, the 4" dimension will help in the setting of partitions as accurately as possible. The approximate dimensions of a 2" × 4" stud is 1½" × 3½". Thus a wall with lath and plaster would be approximately 3½" + ¾" + ¾" or 5" thick. Some architects show stud walls with lath and plaster as 6" (nominal dimension) on the blueprints. If drywall were used the dimensions would be 3½" + ½" + ½" or 4½". Other thicknesses of drywall may be used which would result in different wall thicknesses. Partitions may have to be increased in thickness to take care of special problems. A bathroom partition contains a soil stack which makes it necessary to increase the wall thickness by 2".

Concrete block, tile, or other materials of standard width are usually drawn to the face of the material as shown at G and H whether plastered or not. Solid plaster partitions are located to center lines, see I.

Dimensioning Methods. Fig. 6-12 shows several methods used to draw dimension lines. Most architects use the form shown, with the dimension above the line with slim arrowheads pointing to extension lines which are related to points on the drawings. See J, Fig. 6-12. When the space is too small to permit the dimension to be shown clearly, it may be shown as at K. Some architects use the forms shown at L and M in order to conserve time.

Windows and Doors. The location of windows and doors on floor plans of frame houses is usually determined by measuring to the centers of openings as shown N and O, Fig. 6-13. Windows in brick veneer walls may be located using the center of the opening as at P, Fig. 6-13. Fig. 6-14 (Q, R, and S) shows how windows and doors appear in a masonry wall and how they may be dimensioned. Usually the dimensions are given to the masonry opening.

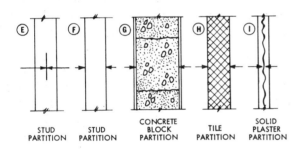

Fig. 6-11

Interior partitions are located by dimensions. Wood stud partitions are located to the center or to the face of the studs. Masonry or tile partitions are located to the face of the material. Solid plastered partitions are located to the centerline.

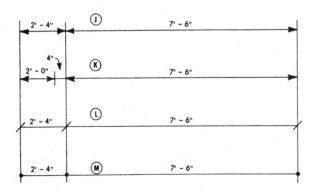

Fig. 6-12

Most architects follow the practice shown by "J" and "K" in showing dimensions.

72 Building Trades Blueprint Reading

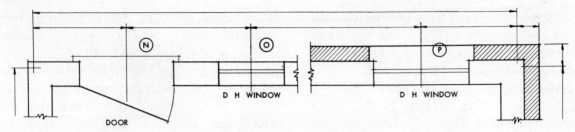

Fig. 6-13 Dimensions are usually given to the centerline of openings in frame and brick veneer walls.

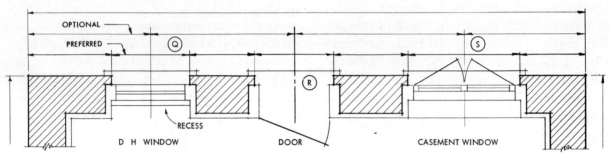

Fig. 6-14 Dimensions are usually given to the masonry opening for brick or other types of masonry. One or the other methods is used.

Scaling and Dimensioning Practices 73

TEST DRAWING NO. 5A (See pages 75-76)

PUBLISHER'S NOTE: THIS DRAWING WAS ORIGINALLY DRAWN TO THE SCALE SHOWN. THE DRAWING WAS REDUCED TO FIT THE PAGE AND CAN NO LONGER BE SCALED.

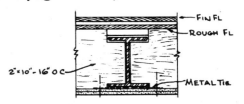

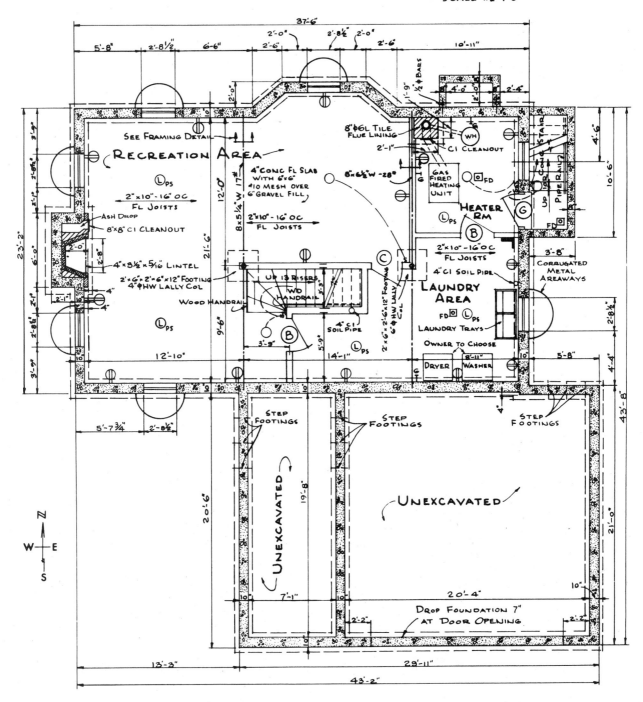

TEST DRAWING NO. 5B (See pages 75-76)

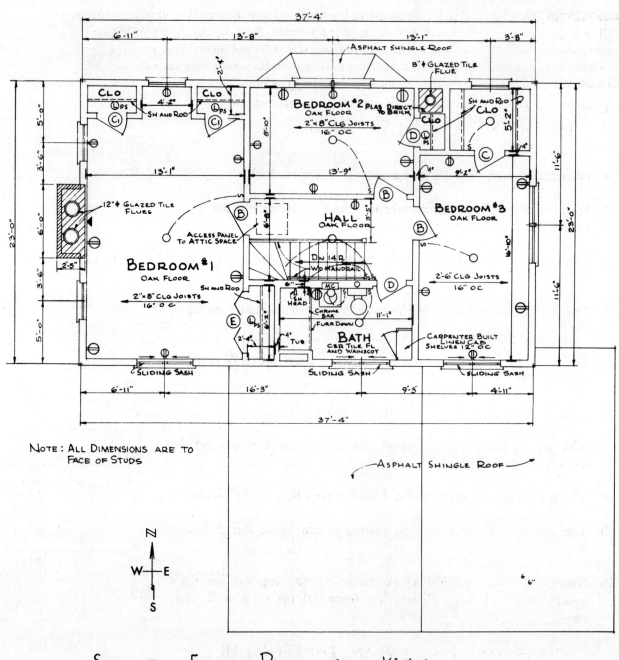

SECOND FLOOR PLAN SCALE ¼"=1'-0"

PUBLISHER'S NOTE: THIS DRAWING WAS ORIGINALLY DRAWN TO THE SCALE SHOWN. THE DRAWING WAS REDUCED TO FIT THE PAGE AND CAN NO LONGER BE SCALED.

Trade Competency Test No. 5

(Chapter 6: Based on Test Drawings No. 5A and 5B)

STUDENT'S NAME   INSTRUCTOR'S NAME

COMPLETION TEST: Key words or numbers have been omitted from each of the following statements. Fill each space so as to make a complete and correct statement. Note: In calculating dimensions of rooms consider the partitions as 4 inches thick and exterior walls as 4 inches thick from the face of the studs. Do not include finish, such as lath and plaster, drywall, wood panelling, etc.
Example: The thickness of the garage foundation wall is <u>10</u> inches.
Appendix A, page 169, is included to provide a review of mathematics.

THE FOUNDATION PLAN, Test Drawing 5A.

For Scoring

1. Foundation walls for the house are _____ inches thick.

2. Foundation walls for the exterior stairs to basement are _____ inches thick.

3. The maximum outside dimensions of the house and garage foundation not including the fireplace and bay are _____ × _____.

4. The outside dimensions of the foundation for the exterior stairs are _____ × _____.

5. The bay foundation is located _____ from the NW outside corner of the house.

6. The overall dimension of the bay (east-west) is _____.

7. The part of the fireplace foundation which projects beyond the outside of the wall is _____ × _____.

8. The hall from the stairs to the laundry area is _____ wide.

9. The distance from center to center of the wide flange beams is _____.

10. Assuming that the wide flange beams supporting the joists are continuous and have 6" bearing (contact) on each wall, they are _____ long.

THE SECOND FLOOR PLAN, Test Drawing 5B.

11. The overall dimensions are 23'-0" × 37'-4". The overall dimensions of the foundation are 23'-2" × 37'-6". The 2 inch difference is for _____.

12. The interior dimensions of bedroom #2 are _____ × _____.

(Cont.)

Building Trades Blueprint Reading

Trade Competency Test No. 5 *(Cont.)*

13. The finished dimensions for the partition (back of the toilet) containing the vent stack is ____ inches. 13._____

14. If the sash of the window in the north wall of bedroom #1 is 3'-0" wide, the window is _____ from the corner of the building (face of the stud corner to near side of window sash). 14._____

15. The masonry dimensions of the fireplace chimney are ____ × ____. 15._____

16. The dimensions of the SE closet in bedroom #1 are ____ × ____. 16._____

17. The chimney flues are ____ inches and ____ inches in diameter. 17._____

18. The east-west dimension of the bathroom is _____. 18._____

19. The interior dimensions of bedroom #3 are ____ × ____. 19._____

20. If the 2" × 8" ceiling joists over bedroom #2 have a full 4 inch bearing (contact) on each partition they are _____ long. 20._____

SCORE: ☐

Tear Off Here

Chapter 7

Reading Blueprints for Structural Information

When an architect draws the plans for a *large* structure, such as an office building or a factory, he has several specially trained men to help him. One of them is an architectural engineer or structural engineer who will draw structural plans to accompany the architectural plans, giving important information about the foundation, the skeleton, and the floor system. The size and location of each steel member will be indicated for each floor; or a complete layout of the reinforcing steel will be drawn if the building is to be made of concrete. Still other drawings called shop drawings will be made by the contractor which will show the men in the shop how the individual pieces are to be prepared for fabrication and how they should be assembled.

The blueprints which are made from the set of structural drawings are the source of information for the craftsman working on a *large* structure. However, this is not the case for the craftsman in the field of *home building*.

In home building the craftsman must learn how to read a set of blueprints and a set of specifications. From this meager information he should be able to visualize the whole framework of the house. See Fig. 7-1. He must be able to lay out the building lines so that the foundation is set in the right place on the lot, and he must determine the grade so that the building will be at the proper height in relation to the established point of measure. His next job is to erect forms so that the foundation, when it is poured, will have the exact dimensions required for the house. He must be able to visualize the supporting members for the floor and to place openings for stairwells and chimneys in the right places. He must plan the walls and partitions in detail, placing the openings in the correct locations and providing for window and door frames with necessary allowances for fitting.

The construction of the roof is usually left to the craftsman with little guidance other than the blueprints of the elevation views and plan views. He must visualize the general shape and then must measure and cut the many members which will serve to support the roof. All of the members used in the building must be made structurally sound. Provisions must be made for wind stress and snow load, depending on the region.

Thus blueprint reading is much more than merely reading the prints. It becomes a matter of interpreting them in terms of members of wood and units of concrete, brick, stone, glass, and other materials not shown in detail on the blueprints.

In order to build a house from a set of blueprints, a great deal of information on the details of construction must be mastered. It is beyond the scope of this book or any one book to give this subject thorough treatment because of the great amount of technical information needed and the many variations between buildings. Many good textbooks are available on carpentry and masonry construction as well as on heating, plumbing, and the

78 Building Trades Blueprint Reading

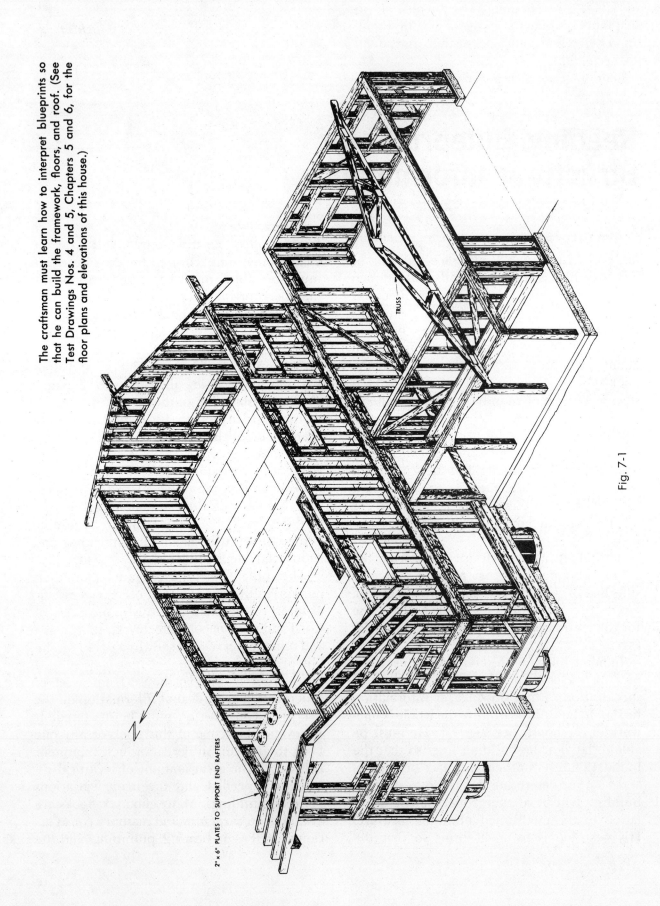

The craftsman must learn how to interpret blueprints so that he can build the framework, floors, and roof. (See Test Drawings Nos. 4 and 5, Chapters 5 and 6, for the floor plans and elevations of this house.)

Fig. 7-1

work of other tradesmen. The purpose of this chapter is to discuss the types of general building construction and some of their important features.

Finding Structural Information on Blueprints. A study of the floor plans and elevation views will tell a great deal about the material to be used in walls, partitions, and floors and where openings are to be placed. The most important clues, however, are found on the blueprints of sectional views. See Fig. 7-2. Views such as this are slices taken vertically through an exterior wall from roof to footing. When the architect uses different material or changes the type of construction, he may show several sectional views taken at different locations. Some of the views will show only the parts which vary from the typical section. For instance, a sectional view may be taken through a fireplace to show construction details, or through a foundation wall to show dimensions.

Construction Types. The three general types of frame construction used today are platform (sometimes called western), balloon, and plank and beam. See Figs. 7-3, 7-6 and 7-7. A masonry veneer building is also considered as frame construction because the internal supporting structure is wood framing, either platform (western) or balloon. See Fig. 7-10.

Solid masonry is used in what is generally known as ordinary construction. See Fig. 7-12. The exterior walls are of brick, stone, concrete block, structural clay tile, or a combination of these units. The other parts of the building, including the floor structure and its supports and the partitions, are of conventional frame construction.

Experiments are going on continually to develop new framing and construction techniques using wood and plywood in many new ways and using steel and concrete in floor and wall systems. Great progress has been made in the use of building components consisting of wall and floor units. Prefabrication has been introduced in which whole houses or parts of houses are built in factories and delivered to the site by truck.

Platform or Western Framing. The most com-

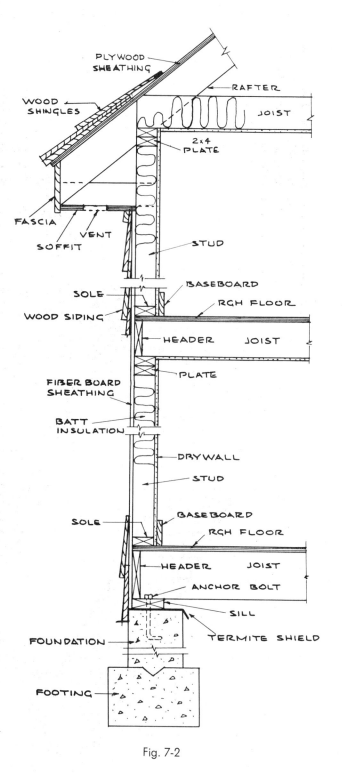

Fig. 7-2

This typical section through an exterior wall of a building with platform (or western) framing shows the important framing members.

mon type of wood framing is called platform (or western) framing. See Figs. 7-2 and 7-3. The main characteristic is that a complete floor system is built as a platform at each level. A flat surface is provided on which to work in fabricating the walls and partitions which are tilted into place when they are completed. See Fig. 7-4. Fig. 7-2 shows a typical sectional view taken through the exterior wall of a building using platform (or western) construction. From information presented in the sectional view the craftsman must visualize how the building is to be constructed in detail, such as is shown in the pictorial view, Fig. 7-3. The many different members in these two figures, except for finished flooring and exterior finish members such as the siding, all have a structural function in the building. Vertical members support floor supporting members and provide strong, rough framed openings for windows and doors. Horizontal members support the floor or roof members. The rough flooring, the exterior wall sheathing, and the diagonal bracing serve to tie the building together so that it can resist the forces which might tend to twist it out of shape. By comparing the two figures (Figs. 7-2 and 7-3), the location and purpose of each member can be determined.

The sill is the lowest wood member of the structure. A termite shield of metal extends over the wall and projects on each side to prevent the invasion of these wood-destroying insects. The sill is anchored to the foundation to maintain perfect alignment and to give strength to the building under wind pressure. The sill is made perfectly level so that the floor joists will all be level. At the center of the building an S-beam girder is located to which a 2 × 4 sill is attached. The girder is supported by posts or columns not shown in either Fig. 7-2 or Fig. 7-3. Joists which support the floor rest on the sills at the wall and at the girder. They are usually spaced on 16 inch centers. A header on the outside wall serves to keep the joists in alignment. Cross bridging placed in rows across the building serves to stiffen the floor. Rough flooring is fastened to the joists. (In this case it is made up of boards laid diagonally to increase the strength of the house. Plywood also serves this purpose very well.) The exterior walls and the partitions are then constructed, tilted

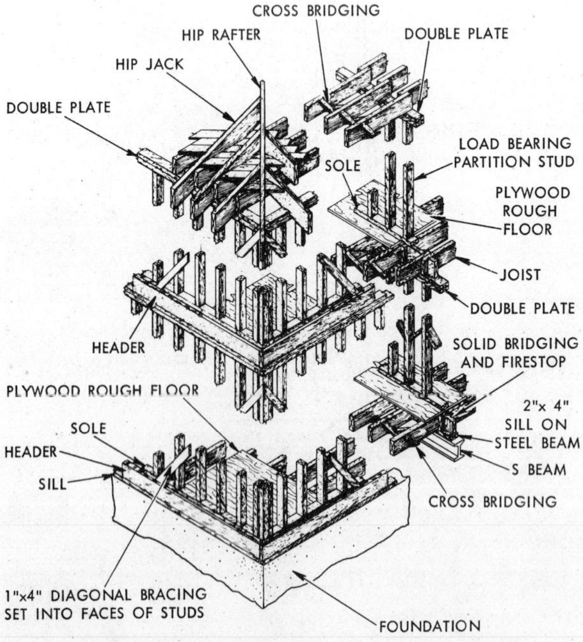

Fig. 7-3

Platform (or western) framing is popular because it can be erected quickly and easily. The rough floor provides a platform for the workman.

Fig. 7-4

Western Pine Association

The rough members which make up the walls and partitions for platform (or western) framing are assembled on the floor and tilted up into place.

up into place, and braced in a vertical position until all of them are erected so that they support one another. Diagonal bracing of 1 × 4 members is often cut into the studs to help the structure withstand wind pressure. (Structural sheathing or plywood sheets, when used instead of sheathing boards, add enough strength so that the bracing is unnecessary.)

If the building is designed to have two floors, as shown in Figs. 7-2 and 7-3, the second floor joists are placed on top of the plates of the first floor walls and bearing partitions and are spiked into place. The rough floor is then laid to cover the whole area. Outside walls and interior partitions are made on the floor and tilted up into position as on the first floor. Ceiling joists are laid on the plates. Rafters are cut so as to bear on the wall plates with a flat surface sufficiently large in area to transfer the load of the roof to the wall.

If the building has only one floor, ceiling joists are erected to rest on the plates of the first floor walls and bearing partitions. Roof rafters are cut so as to bear on the wall plates.

The matter of shrinkage of wood members in a frame building is quite important. Lumber will shrink across the grain, that is, across the width of the plank, enough to make it a factor to consider. The shrinkage lengthwise is negligible. In the platform (or western) type of framing, the shrinkage of wood members placed on edge in the outside wall and those used at the center of the building is equal because members of the same dimension are used in both places.

Balloon Framing. The distinctive feature of balloon frame construction is that the studs in the outside wall are full length from the sill to the upper plate. The studs at the bearing partition are made as long lengths as conveniently possible. Thus shrinkage is almost eliminated. (This is a particularly valuable feature when brick veneer or stucco is used on a building of more than one floor in height. The wood part of the wall and the veneer part will remain in horizontal alignment.) Figs. 7-5 and 7-6 are typical of balloon framing. These figures show the first floor joists extend-

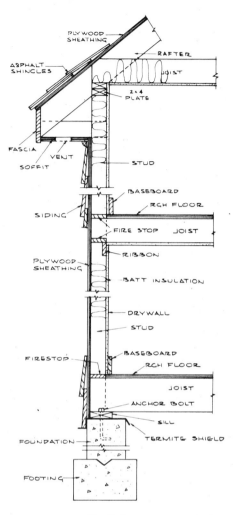

Fig. 7-5

This typical section through an exterior wall of a building with balloon framing shows the distinctive features of long studs, ribbon, and firestops.

ing to the outer face of the studs, thus eliminating the header. At the second floor level the joists rest on a ribbon which is a narrow board cut into the studs. Joists are also spiked into the studs. The details of construction at the eaves are the same as that for platform (or western) framing.

It is important that firestops (or draftstops) be installed between the studs as shown in Fig. 7-5. These short pieces are carefully fitted into the space between the studs and serve to retard the spread of fire. If they were not used, a fire could easily travel in the space in the wall from basement to roof or between the floor and ceiling at each floor level.

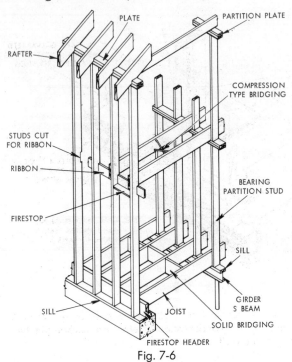

Fig. 7-6

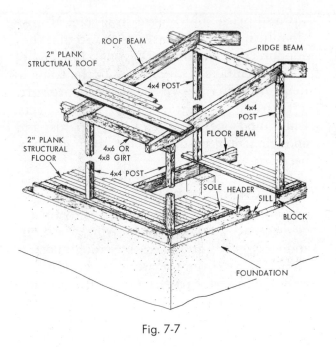

Fig. 7-7

Plank and beam framing requires the use of heavy structural members and planks for floors and roofs.

In balloon framing the studs extend from the sill to the plate for the rafters. The ends of the second floor joists rest on a ribbon.

Plank and Beam Framing. Plank and beam framing is used extensively in one story houses because it permits the use of wide expanses of glass in exterior walls and provides a novel effect inside with heavy exposed beams and plank ceilings. See Figs. 7-7 and 7-8. The beams used to support the floor and roof are spaced at wide intervals. Either solid or built-up members are used.

Planks nominally 2 inches thick are required for floors and roofs in order to span the distance and have structural value. (A plank with a *nominal* 2" thickness would have an *actual* thickness of 1½" after it is dressed. A 2 × 6 plank, for example, would be 1½" × 5½".) Tongue and grooved planks which show an interesting "V" groove on the bottom are often used when planks are exposed, Fig. 7-9.

Posts are spaced at equal intervals in the outside wall to support the ceiling beams. Posts may be used at the center of the building or a partition may be used to support the ridge beam. When posts are used in the exterior wall and also at the center of the building, the construction may be called "post and beam construction," although "plank and beam" is the more common description.

Weyerhaeuser Company

Fig. 7-8

Plank and beam framing gives a special effect of exposed heavy beams and high sloped ceiling. The supporting posts are uniformly spaced in the walls.

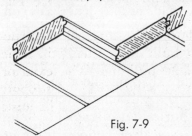

Fig. 7-9

Tongue and grooved planks are used when the planks also serve as a ceiling.

Reading Blueprints for Structural Information 83

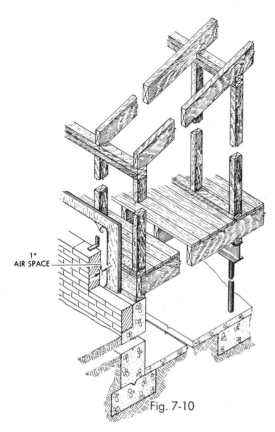

Fig. 7-10

A brick veneer house is a frame house with a skin of brick. A 1 inch space permits a slight amount of movement and provides an air space.

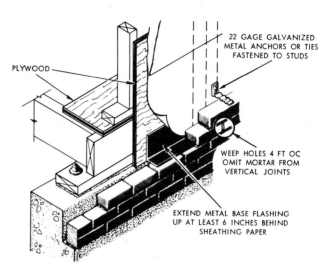

Fig. 7-11

Special precautions must be used in a brick veneer wall to provide for condensation. Metal ties anchor the two parts of the wall together.

Brick Veneer Construction. Brick veneer construction is actually frame construction with a layer of brick or other masonry units making up the outer surface. Fig. 7-10 shows how a brick veneer house is constructed using balloon type framing. A 1-inch air space is maintained between the brick and the sheathing to permit a slight amount of shifting between the two parts of the wall and to provide an air space. This space is kept open by means of weep holes to permit the escape of condensed moisture. The two parts of the wall are tied together with metal ties. See Fig. 7-11.

Masonry Buildings. The internal construction of a masonry building is usually the same as that for a frame building. The exterior walls however are of solid brick and support the ends of the joists. See Fig. 7-12. The joists rest on a thickness of brick (a wythe) as shown at the first floor level or rest in the brick wall as shown at the second floor level.

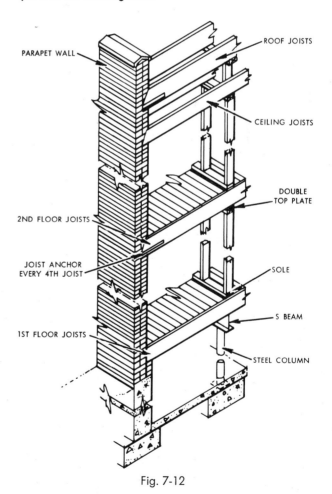

Fig. 7-12

Typical ordinary construction for a masonry building uses wood joists and wood bearing partitions internally.

84 Building Trades Blueprint Reading

Fig. 7-12. Fig. 7-13 shows how the ends of the joists are cut so that in the case of a severe fire, in which the joists burn through and fall, they do not force the wall to fall outward.

Furring strips are placed on the inside of the wall to provide a level nailing base for the lath or drywall. See Fig. 7-14.

Buildings Using Unit Components. Many experiments have been made to develop building component parts so that they may be built on the job or away from the job in a shop. See Figs. 7-14 to 7-16. The panels consist of straight wall units, units with doors or windows, gable end units, etc. See Fig. 7-16. Flexibility is obtained so that plans may vary greatly.

Prefabricated Houses. Buildings made in factories have been developed successfully in many parts of the country. Some buildings consist of parts such as floor and roof trusses, and walls or partitions made in sections. Figure 7-14 shows common roof and floor trusses.

The wall sections in some cases are made complete with interior and exterior finish as well as contained electric wiring. Other wall units contain windows and doors. See Figs. 7-15 and 7-16. The units are delivered to the job, are erected and fastened to the floor and to each other to make the complete house structure. The units are called "components" and the method of building is called Panelized Construction.

Another type of prefabricated construction is called Modular Construction. Whole rooms or parts of buildings are made in factories and delivered to the job to be lowered to the foundation. The units are called "modules". Some of them are complete kitchens or bathrooms including fixtures, equipment, wiring and plumbing.

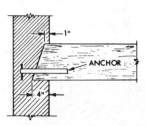

Fig. 7-13

The fire cut on joists permits the joists to fall without destroying the wall.

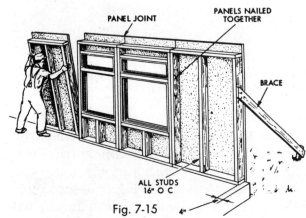

Fig. 7-15

Components may be made in the shop or on the job.

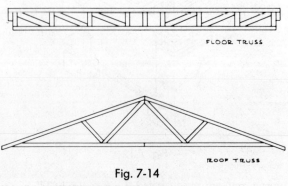

Fig. 7-14

Common roof and floor trusses.

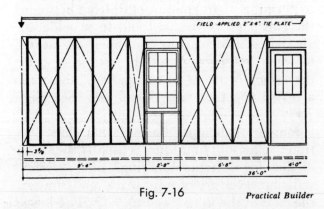

Fig. 7-16 *Practical Builder*

Panels are made in units which are multiples of 4 inches. The Unicom method of house construction was developed by the National Forest Products Association.

Self-Check Quiz No. 7

(Based on Fig. A)

Identify the various parts by using the correct number. *Answers are given in the appendix.*

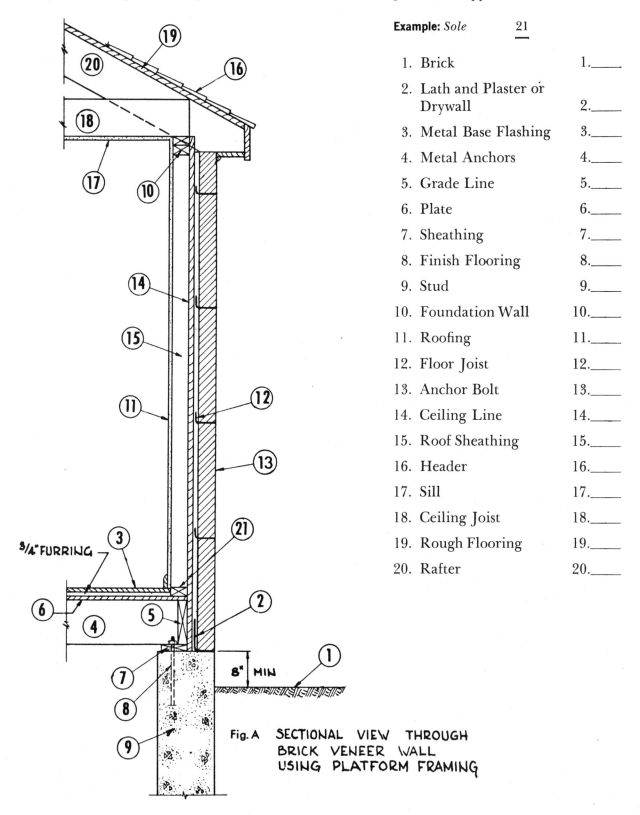

Fig. A SECTIONAL VIEW THROUGH BRICK VENEER WALL USING PLATFORM FRAMING

Example: Sole   21

1. Brick   1.____
2. Lath and Plaster or Drywall   2.____
3. Metal Base Flashing   3.____
4. Metal Anchors   4.____
5. Grade Line   5.____
6. Plate   6.____
7. Sheathing   7.____
8. Finish Flooring   8.____
9. Stud   9.____
10. Foundation Wall   10.____
11. Roofing   11.____
12. Floor Joist   12.____
13. Anchor Bolt   13.____
14. Ceiling Line   14.____
15. Roof Sheathing   15.____
16. Header   16.____
17. Sill   17.____
18. Ceiling Joist   18.____
19. Rough Flooring   19.____
20. Rafter   20.____

TEST DRAWING NO. 6 (See pages 87-88)

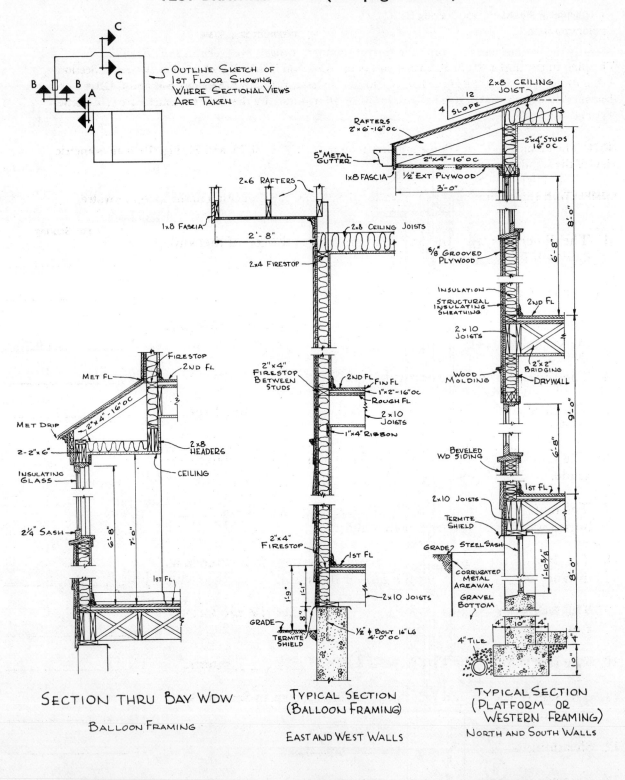

Trade Competency Test No. 6

(Chapter 7: Based on Test Drawing No. 6)

STUDENT'S NAME _____   INSTRUCTOR'S NAME _____

The plan in the upper left hand corner shows the direction in which each section is viewed. Section A-A is drawn as though the house had platform (or western) framing. Sections B-B and C-C are drawn as though the house had balloon framing. Dimensions are the same for similar parts of Section A-A and B-B.

NOTE: The blueprints of this house are shown on pages 27, 28, 60, 73, and 74. Fig. 7-1 is an isometric view of the framing.

COMPLETION TEST: Fill in the missing words, dimensions, or information in the spaces provided.

For Scoring

1. The dimension from the finished first floor to the top of the window sash is _____.   1. _____

2. The ceiling height in the bay is _____.   2. _____

3. The distance from the top of the basement floor to the ceiling of the second floor is _____.   3. _____

4. The thickness of the foundation wall is _____.   4. _____

5. The finished first floor in Section B-B is _____ above the grade.   5. _____

6. The lowest wood member in Section B-B is _____ above the grade.   6. _____

7. The roof rafters shown in Section A-A extend a distance of _____ horizontally from the face of the studs.   7. _____

8. The face of the outside rafter shown in Section B-B extends a distance of _____ from the face of the studs.   8. _____

9. The roof slope, Section A-A, is _____ inches rise per foot of run.   9. _____

10. The roof slope of the bay (is or is not) _____ shown.   10. _____

11. The sash of the 1st and 2nd floor windows shown in Section A-A are made of _____.   11. _____

12. Sheathing is _____.   12. _____

13. Exterior finish (exterior wall covering) is _____ and _____.   13. _____

(Cont.)

Tear Off Here

Trade Competency Test No. 6 *(Cont.)*

14. The interior wall finish is _____.   14. _____

15. Studs are ____ × ____ spaced _____.   15. _____

16. Floor joists are ____ × _____.   16. _____

17. Second floor ceiling joists are ____ × ____.   17. _____

18. Rafters are ____ × ____.   18. _____

19. Mineral wool batt insulation is shown in the exterior walls. Where else is it shown? _____
    19. _____

20. In the place of a gutter, a _____ is shown over the bay.   20. _____

21. What distinctive framing members are shown in Section B-B (balloon framing) which are not shown in Section A-A (platform or western framing)? _____ and _____   21. _____

22. Where is a termite shield shown in Section B-B? _____
_____   22. _____

23. What does the designation "½"⌀ bolt 16" lg 4'-0" o c" mean? _____   23. _____

24. What is the purpose of the cross bridging shown in Section A-A? _____   24. _____

25. What is the purpose of the 2 × 8 inch headers in Section C-C? _____   25. _____

SCORE: ☐

Tear Off Here

Chapter 8

Reading Detail Drawings

The architect tries to show, on the plan views and the elevation views, all of the graphical information required to build a building. (Specifications give additional information in the form of written explanations and descriptions.) However, almost every time the architect draws a set of plans, he finds it necessary to show some part which does not ordinarily appear on the plans and elevations in detail. He finds it necessary to show some features at a larger scale because enough information cannot be crowded into the space of the small scale drawings. Some of the details are sectional views taken through a part of the building, such as a fireplace or a foundation wall, to show construction and dimensions.

When an architect prepares drawings for a *large building*, he assembles what is called an architectural set of plans which contains floor plans, elevations, section drawings, and as many details as necessary to enable the various contractors to prepare estimates. After the contracts are let, detail drawings, often numbering several hundred, are prepared to show items of construction, elevations of rooms, full size details of ornamentation and trim, etc. It is understood that the details may not alter the basic information given on the architect's original plans. Their purpose is to clarify the many things which could not be drawn earlier because of lack of time.

The types of details which are shown on a set of blueprints of a *small home* generally fall into these classes:

A. Elevations of kitchens, bathrooms, and walls in other rooms which have some special features. These are generally drawn at the same scale as the floor plans unless they are very complex.

B. Sectional views through parts of the building, such as through a fireplace or through the foundation at various places. Structural information to show how floor members should be placed at a stairwell or to show how a truss is to be constructed.

C. Details of windows and doors. Views showing how special frames are to be made for bay windows or other unusual arrangements of windows.

D. Details of exterior trim such as cornices, dormers, front entrance doorways, etc. Details of interior trim such as built-in cabinets, mantels, paneling, and trim.

The Scale Used for Details. Detail views may be drawn at the same scale as the plan views (usually $\frac{1}{4}'' = 1'\text{-}0''$) if they are not too complex. The use of the same scale is particularly applicable to the drawing of room elevations. This allows dimensions to be transferred directly from plan views to the elevations. Construction details are usually drawn at a larger scale so that they can be shown with greater clarity.

The architect may choose one of several scales, depending primarily on how large the detail must be drawn in order to show all of the information clearly. A second consideration, of course, is the amount of space available on the sheet. The scales preferred are:

$\frac{3}{4}'' = 1'\text{-}0''$ $3'' = 1'\text{-}0''$
$1\frac{1}{2}'' = 1'\text{-}0''$ FULL SIZE

Scales of $\frac{1}{2}'' = 1'\text{-}0''$ and $1'' = 1'\text{-}0''$ are not considered the best practice.

The reason for using the scales suggested above is that the builder who is to use the blueprints carries a folding rule divided into inches and sixteenths of an inch. If the detail were drawn at the scale of $\frac{3}{4}'' = 1'\text{-}0''$, each $\frac{1}{16}$ inch on his rule would represent one inch. If the detail were drawn at a scale of $1\frac{1}{2}'' = 1'\text{-}0''$, each $\frac{1}{8}$ inch would represent one inch. If the detail were drawn at the scale of $3'' = 1'\text{-}0''$, each $\frac{1}{4}$ inch on his rule would represent one inch and each $\frac{1}{16}$ inch would represent $\frac{1}{4}$ inch. A drawing made at the scale of $3'' = 1'\text{-}0''$ is one fourth actual size. Full size details are drawn when the exact shape of a molding, parts of a curved stair rail, etc., are required.

Dimensions. Dimensions are shown on detail views if they are not shown elsewhere on the plans and elevations or if they add greatly to the understanding of the view. It is important that there be control over dimensioning in general. When the same dimension is shown on two different views, confusion would result if one were changed and the other were not. When dimensions are shown only once, it may be necessary to refer to several blueprints in order to find a desired dimension.

Locating the Detail. In many instances it is necessary to show the exact place where the section view is taken and the direction in which the imaginary slice is viewed. This is accomplished by drawing a cutting plane line through the plan or elevation view with two arrows showing the direction the reader is to look to see the detail. See Fig. 8-1.

A notation SEE DETAIL is frequently used to call attention to the fact that more information is obtainable about the designated part of the building somewhere else on the same sheet or on another sheet of the set of blueprints. Wherever there are several details, they will be designated DETAIL 1, DETAIL 2, etc. When several sectional views are taken through walls, foundations, or other parts of the structure, the section drawings are designated SECTION A-A, SECTION B-B, etc.

Whenever a detail is drawn it should be

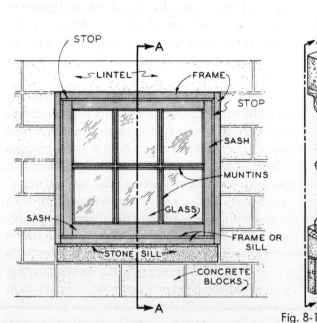
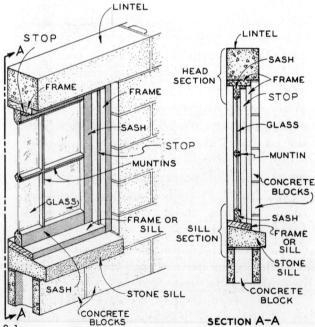

Fig. 8-1

Section line A-A shows where an imaginary slice is taken through a basement window in a concrete block wall. Arrows indicate the direction in which the view is taken.

The section view at the right, SECTION A-A, shows each part of the window and frame in detail.

placed on the same drawing where reference is made to it, if at all possible. Occasionally all of the details are gathered together on one sheet.

Details of Interior Wall Elevations. The plan view of a kitchen shows the location of cabinets, sink, dishwasher, and all permanent built-in features. It also shows where space is provided for movable equipment such as the refrigerator and range. This information is not sufficient for the millman who makes the cabinets or for the carpenter who must build the soffit over the cabinets. (A soffit in this sense is a lowered ceiling.) Elevation views are necessary to show the arrangement of cabinet doors and drawers and to give information about soffits, lighting, hoods, fans, and many other features.

Cabinets are often built to order to fit perfectly into the space provided. Dimensions of the cabinets, except for height dimensions, are not usually shown on the detail elevation views. However, the millman measures the space carefully and then proceeds to make the cabinets to fit. He studies the detail drawings in order to use the desired arrangement of doors and drawers. He makes the counter top, cutting out the pieces, where the sink and range top will fall into place. When stock cabinets are used, adjustments are made by the use of fillers so that cabinets will fill the whole wall space. Fig. 8-2 shows a typical modern kitchen with a counter top range and a built-in oven. Fig. 8-3 shows how the detailed elevation views of a kitchen look on a blueprint.

Bathroom elevation views are generally shown in detail also. The location of the fixtures is obtained from the plan view. The plan view is inadequate to show all of the features of the bathroom unless elevation views or extensive notes are provided. The elevations tell the prospective owner at a glance how the room will look. They will give the builder important information about features, such as a dropped ceiling over the tub. They will also indicate details about drawers, the mirror, and lighting for a cabinet-lavatory. They may show the location of accessories such as towel bars, soap dishes, etc. See Fig. 8-4.

Many times the walls of other rooms have special treatments which would make it advisable to draw them in detail. Such instances might be wood paneling and fireplace arrangements, built-in features such as chests of drawers in dining rooms, or book shelves in library or living room.

Details of the Structure. Chapter 7 was devoted to reading blueprints to find structural information and to interpreting the vertical section through the wall. There are occasions when more structural information than this one main sectional view would be useful.

It occurs frequently that a foundation wall must be formed differently at various places on the perimeter of the building to meet special conditions. Several detailed sections are drawn, each labeled with corresponding letters and with arrows to show the direction of the cut. See Fig. 8-5.

Fireplace details cannot be shown adequately on plan views and small scale elevation views. The design demands careful at-

Fig. 8-2 *Kitchen Kompact*
Counter top ranges and built-in ovens are features of modern kitchens.

92 Building Trades Blueprint Reading

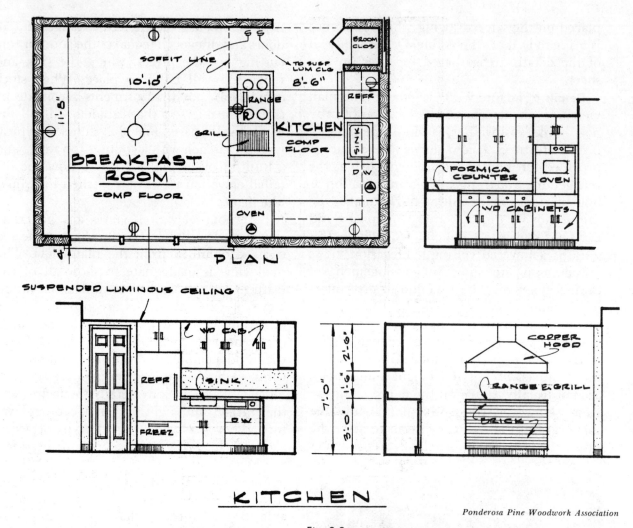

Fig. 8-3

Detail views of kitchen elevations supplement the information shown on the plan view.

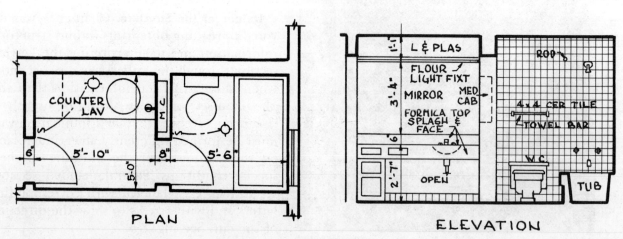

Fig. 8-4

Detailed elevations of bathrooms show information on wall treatment, cabinets, lighting, and accessories.

Reading Detail Drawings

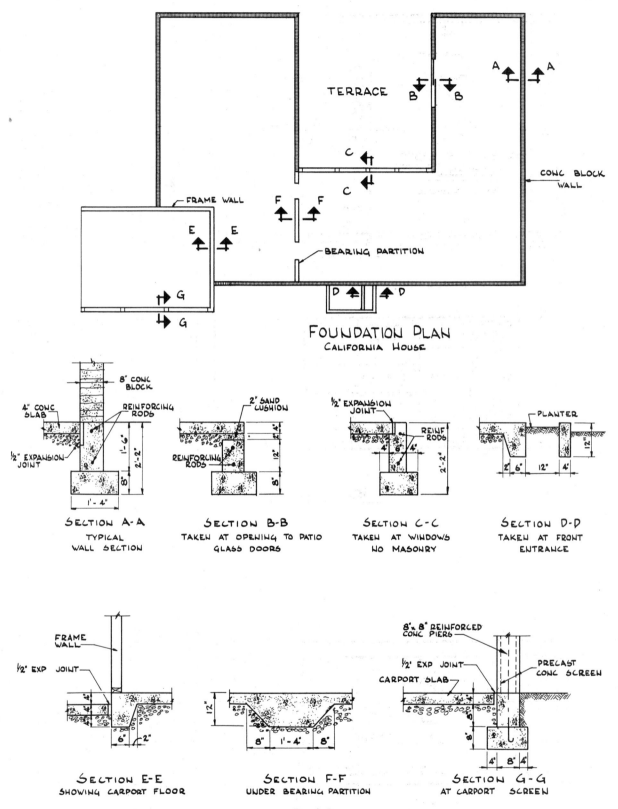

Fig. 8-5

Detail drawings of sectional views taken at various places in the foundation wall will give all of the information needed to build the form work.

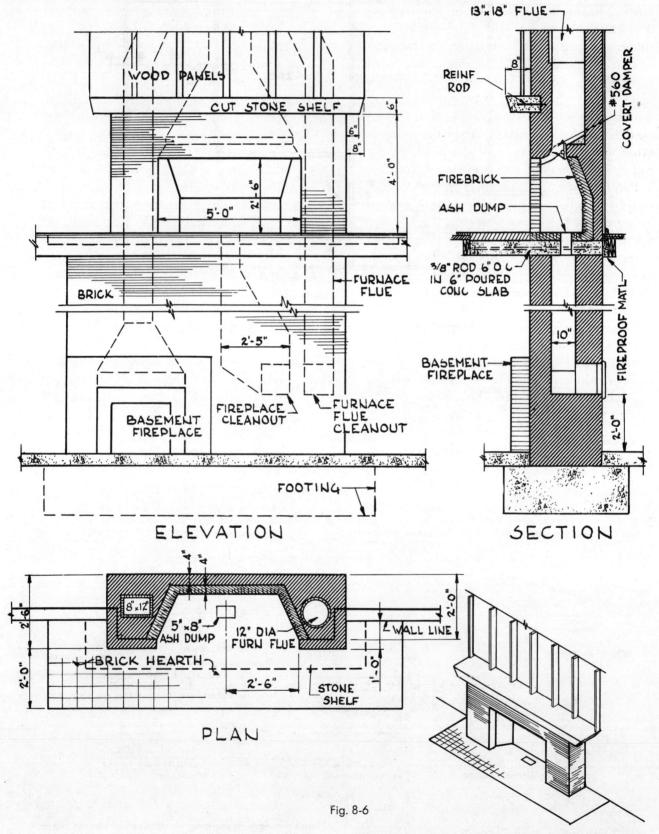

Fig. 8-6

Detail views of a fireplace are necessary to show information about construction, flues, dimensions, etc.

tention to the shape and size of the fireplace, the arrangement of the damper, and the shape and size of the transition to the chimney (the throat). The mason must follow prescribed dimensions in order that the fireplace have efficient combustion and draft. The hearth must be designed so that there is no danger of transmitting heat to floor joists; other fire precautions must also be made. See Fig. 8-6 for views of a conventional fireplace.

Patented metal forms may be used which simplify the building of a fireplace. The form becomes the back and sides of the fireplace opening and contains the damper and throat. A sectional view is still advisable because the masonry which is to be built around the form and the structural support for the hearth must be shown.

The architect usually leaves details of openings in walls and floors and other matters having to do with rough framing to the builder. However, where structural design is involved and loads must be supported in a certain way, he may give a detailed view to show how he wants the job done. This may include details of how an opening for a stairwell should be framed or how walls or floors should be constructed to minimize sound transmission.

Roof framing plans are rarely shown unless there is a very complex problem to solve. Trusses are often detailed, see Fig. 8-7, because the safe load characteristics are the responsibility of the architect.

Window and Door Details. Window and door details have always been included on blueprints until recent years. Today, however, doors and windows are standardized in sizes and types and are sold "packaged" so that the windows, doors, and frames come to the job ready to be installed in the wall. Wherever an unusual situation arises, however, it is necessary to make a detail to bring out the structural support, the manner in which the window or group of windows are to be fabricated, how they are to be installed, and how the trim is to be applied. Such a situation occurs when there is a bay window or a combination of several windows. An understanding of the various parts of a window and how they fit together should be part of the background of the building craftsman.

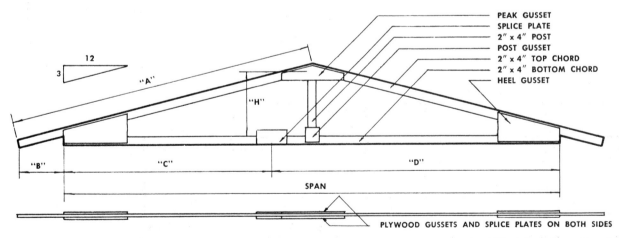

University of Illinois Small Homes Council

Fig. 8-7

Details of a king post, nail-glued roof truss using 2 × 4 members are shown. A typical application would be:

Span = 22'-0"
A = 14'-0"
B = 30⅝"
C = 9'-0"
D = 13'-0"
H = 33"

96 Building Trades Blueprint Reading

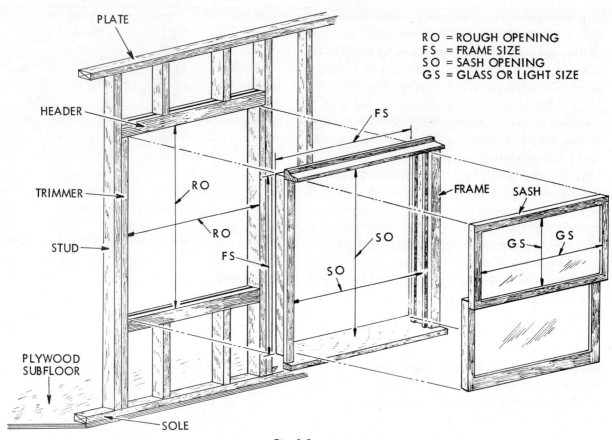

Fig. 8-8

This simplified drawing shows how the sash fits into the frame and how the frame fits into the rough opening.

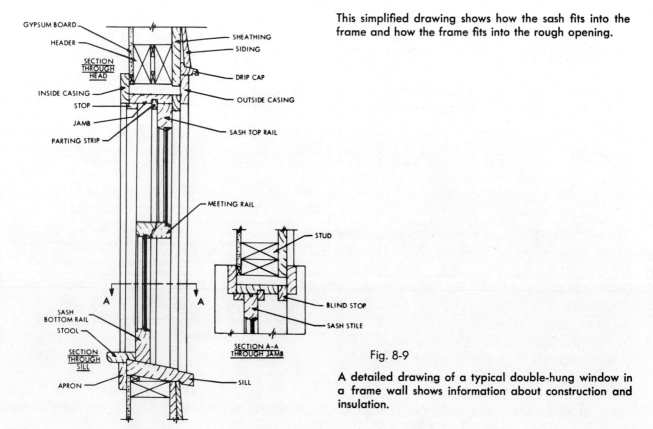

Fig. 8-9

A detailed drawing of a typical double-hung window in a frame wall shows information about construction and insulation.

Fig. 8-8 shows how the sash fits into the frame and how the frame fits into the wall. Fig. 8-9 shows a sectional view of a double-hung window in a frame wall with all the parts shown in detail. Section A-A, taken through the jamb, shows detail of what would be seen if the side of the window were cut by a cutting plane. There is much similarity between the section through the jamb and the section through the head. However, there is a difference in the rough structural members (shown with a cross); the drip cap is only shown in the head; the lower sash is shown in the section through the jamb; and the upper sash in the section through the head. Otherwise the views are identical. Self-Check Quiz No. 8 shows the same window in isometric.

Fig. 8-10 shows an outswinging awning type window which is delivered to the job with the glass in place, the window in the frame, and with all hardware attached and ready to go. Fig. 8-11 shows a detail of this window with dimensional allowances so that the builder can make the opening the right size. In Fig. 8-11 the section through the jamb is revolved 90° and placed in line with the section through the head and sill. The section through the head and the section through the jamb show the difference between the side and top pieces of the sash and the side and top pieces of the window frame.

NOTE: If you cut a piece of paper or filing card about 1½ x 3½ inches, you can cover up the *Section Through Jamb* in Fig. 8-11. You will then see the window cut by a vertical plane with the Head and Sill sections in proper relationships.

Details of Exterior and Interior Trim. Details of exterior trim are often worked out in large scale views so that the builder may know exactly what the architect wants. The cornice is one of the places where such a detail may be found useful. See Fig. 8-12. An unusual use of siding, a dormer, a traditional or contemporary front entrance doorway, or a particular pattern of laying brick for ornamental purposes may be detailed.

Interior finish includes all of the special items of cabinet and millwork and wall finish.

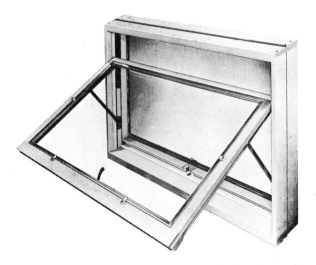

Andersen Corporation

Fig. 8-10

An outswinging awning type window is delivered with frame, sash, weather stripping, and hardware.

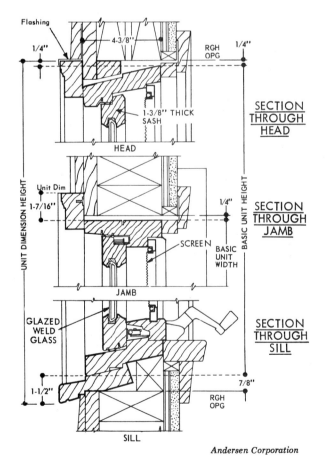

Andersen Corporation

Fig. 8-11

This section gives the details of an awning type window.

98   Building Trades Blueprint Reading

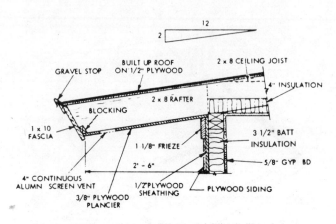

Fig. 8-12

The detail of the cornice shows the builder how to prepare and assemble the framing and how to trim members.

Southern Pine Association

Fig. 8-13

A built-in closet wall serves as a room divider. The closet wall will require a detail view. (Note the plank and beam construction.)

Cabinets in kitchen and bathrooms, built-in wardrobes, book shelves, and free-standing room dividers require detailing because they are built to order. See Fig. 8-13.

Stairs are generally built by a carpenter or are assembled from stock parts. When stairs

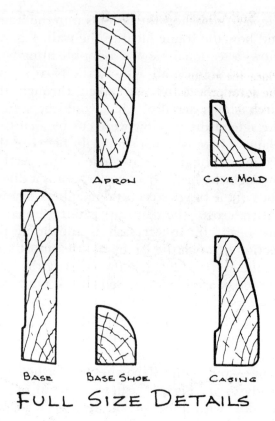

FULL SIZE DETAILS

Fig. 8-14

Moldings are shown full size so that the millman will have the exact profile.

are ornamental or present some particular problem which needs explanation, detail views are made.

The trim used in a house such as the wood base, cornice, window casing, and chair rail are usually stock material, purchased at a lumber yard. An architect may want to use moldings, however, which are made especially for the job if he is designing a traditional house or an ultra-modern house. He will then show cross section views at full size on his drawings so that the men in the mill who make the molding may prepare their knives to cut the exact profiles. See Fig. 8-14.

Casing is the wood trim around doors and windows. An apron is placed under the interior window stool. Cornice cove is used at the top of the wall where it touches the ceiling, however, it is used very seldom nowadays.

Self-Check Quiz No. 8

Place the name of the part in the space provided. *Answers are given in the appendix.*

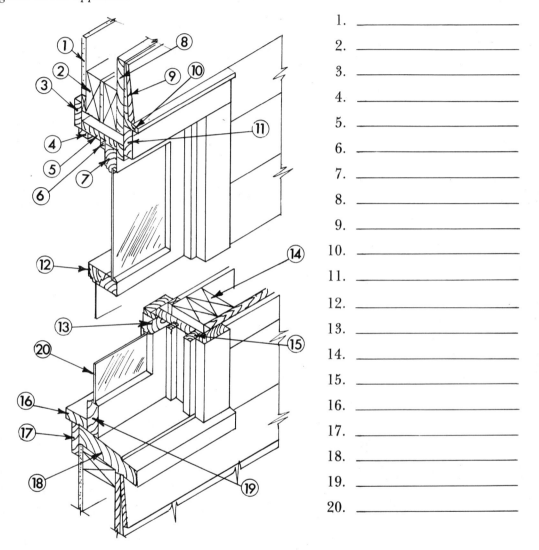

1. _____
2. _____
3. _____
4. _____
5. _____
6. _____
7. _____
8. _____
9. _____
10. _____
11. _____
12. _____
13. _____
14. _____
15. _____
16. _____
17. _____
18. _____
19. _____
20. _____

TEST DRAWING NO. 7 (See pages 101-102)

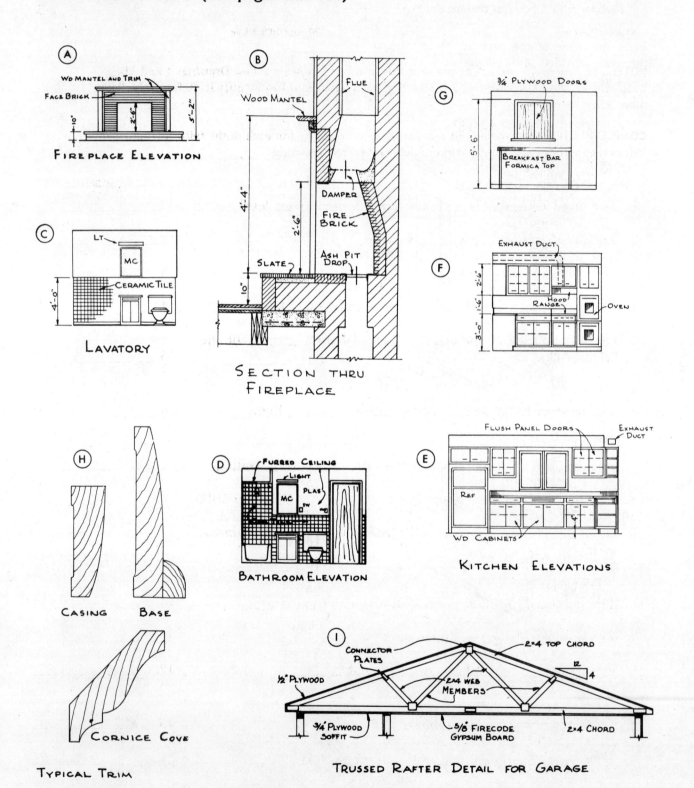

Reading Detail Drawings

Trade Competency Test No. 7
(Chapter 8: Based on Test Drawing No. 7)

STUDENT'S NAME

INSTRUCTOR'S NAME

NOTE: The first and second floor plans of this house are shown on Test Drawings 4 and 5B. Each view on Test Drawing No. 7 has been given a letter A to I to identify it. Refer to the floor plans when necessary.

COMPLETION TEST: Key words and numbers have been omitted in each of the following statements. Fill in the spaces so that they make complete and correct sentences.

DETAILS A AND B

For Scoring

1. The fireplace hearth is ____ above the finished floor level.   1._____

2. The fireplace opening is ____ high.   2._____

3. The fireplace hearth is supported by a slab of _____.   3._____

4. The chimney is made of _____ and the mantel of _____.   4._____

5. Firebrick is used at the _____ and on the _____ of the fireplace.   5._____

DETAIL C

6. The lavatory has a ceramic tile wainscot _____ high.   6._____

DETAIL D

7. The dropped ceiling over the tub is called a _____ ceiling.   7._____

8. The end of the tub enclosure has a _____ wall finish.   8._____

9. Other parts of the bathroom walls have a _____ wainscot with _____ above.   9._____

DETAIL E

10. The number of cabinet doors shown in elevation is ____.   10._____

11. The number of drawers shown in elevation is ____.   11._____

12. Ventilation is provided by _____ and _____.   12._____

DETAIL F

13. The counter is ____ above the floor.   13._____

14. The soffit above the cabinet is ____ above the floor.   14._____

15. Part of an upper cabinet is sacrificed because of the _____.   15._____

(Cont.)

Tear Off Here

Trade Competency Test No. 7 *(Cont.)*

DETAIL G
16. The blank space at the right is because of the _____. 16._____

DETAIL H
17. The casing is a trim member used around _____ and _____. 17._____

18. The cornice mold is used on living room walls at the _____. 18._____

19. Typical trim details such as in Detail H are usually shown _____ size. 19._____

DETAIL I
20. The trussed rafter is used in the building to support the _____ roof. 20._____

21. The main wood members used are ____ × ____. 21._____

22. The roof sheathing is _____. 22._____

23. The slope of the truss is ____ inches in 12 inches. 23._____

24. The plywood soffit is over the _____. 24._____

25. The garage ceiling is _____. 25._____

SCORE: ☐

Chapter 9

The Plot Plan

The plot plan is usually shown on the first sheet of a set of blueprints. It shows the shape of the lot on which the building is to be built, the lot dimensions, and the angles at each corner. The dimensions of each building already built on the lot and those to be built are shown, and dimensions are given locating them in relation to the property lines. The location of walks, drives, courts, patios, etc., are given. An arrow indicating the north point is included. Utilities such as gas mains and electric power source are shown. Water mains, and sewers are identified with a line symbol. The relationship of levels (elevations) at various points on the lot and the first floor level (elevation) of new buildings are indicated. Contour lines show the *natural* slope of the ground. A second set of contour lines shows the slope *after* the fill has been redistributed and the construction is completed. Existing trees and shrubs are usually shown, and those which are to be removed are designated.

The local building authorities require that a plot plan be submitted to them so that they may determine quickly whether or not the ratio of the area of the house to the area of the lot, and the location of the house on the lot, follow the building ordinances. For instance, some communities do not permit a single family dwelling to cover more than 30 per cent of the area of the lot. Also, the distance from the side building line to the property line may not be less than five feet. Other cities may require this distance to be 10 per cent of the width of the lot. Buildings in a block must remain back of a building line established by the city. See Fig. 9-1. The building on the corner lot has an area of 1440 square feet not including the garage. The lot has an area of 7200 square feet. Thus the house only covers 20% of the area of the lot. This should be quite satisfactory. Dimensions A and B (see Fig. 9-1) from property lines to building lines are usually established by the city. Minimum dimensions C and D between houses are often established but may vary depending on the type of structure, number of floors in the houses, etc. Minimum side limits E and F are specified regardless of other factors. Minimum side and rear yard limits H and G may be set also. Each town and city has its own requirements regarding area and placement limits.

Restrictions are placed on buildings in relation to the lots on which they are to be built in order:

> To provide for (a) convenient access to and circulation around the dwelling, (b) adequate natural light and ventilation of rooms and spaces, (c) reasonable privacy for each living unit, (d) utilization of the plot for laundry drying, gardening, landscaping and outdoor living, (e) and where individual water-supply and sewage-disposal systems are involved, adequate areas to assure a safe and sanitary installation.[1]

The Survey Plat. The basic information for the plot plan is found on a survey plat which

1. *Minimum Property Standards for One and Two Living Units,* F H A No. 300, Section 500.

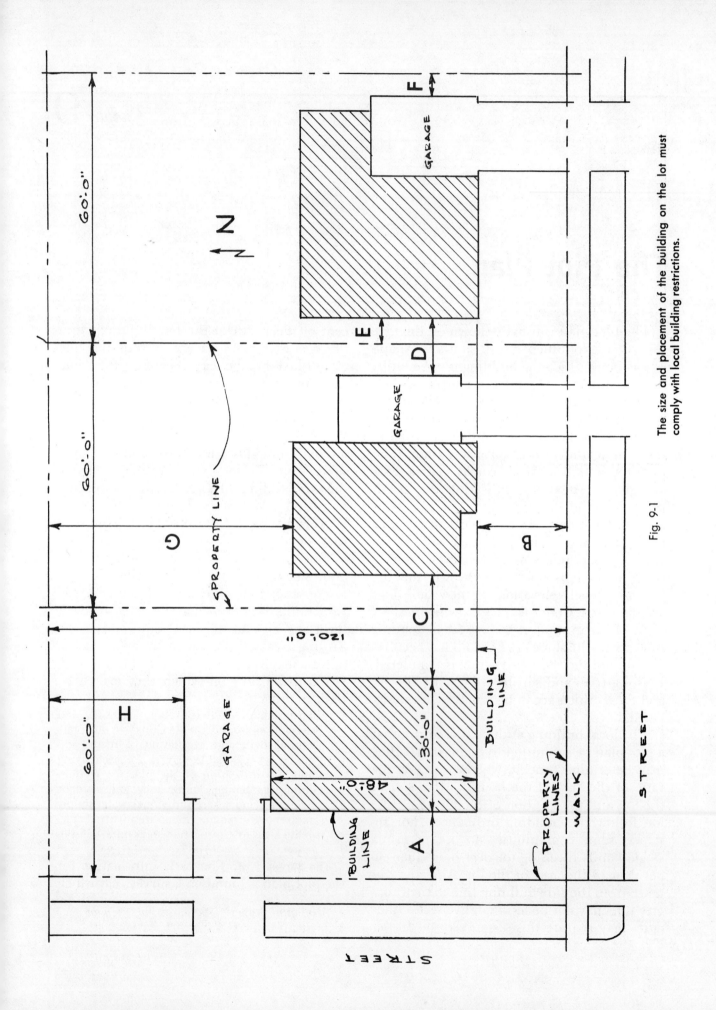

Fig. 9-1 The size and placement of the building on the lot must comply with local building restrictions.

The Plot Plan

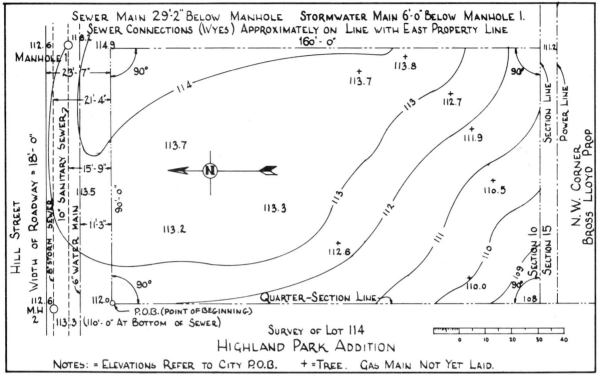

Fig. 9-2

A survey gives the exact shape and size of a lot in relation to a "point of beginning." The natural grade is shown by contour lines at 1'-0" intervals in this case.

is drawn by a licensed surveyor. He obtains the legal descriptions of the piece of ground from the deed to the property which he receives from the owner. Using his surveying instruments and a tape or chain he locates a corner of the property in reference to a *Datum Point* which has been established by the town, city, or county. This is usually a marker made of concrete and embedded in the ground, or it may be any other point which is designated by ordinance to serve the purpose. In a city, the marker may be placed on a street or a walk. It could even be a notch on the corner of a building. When a new subdivision is opened the surveyor may have to establish accurate direction lines, measure precise distances, and work out differences in level (elevation) from a *Datum Point* which is far away. After the surveyor locates a corner of the property or some central point, a stake is driven firmly in place or a concrete bench mark is set. This in turn becomes the point of reference for the location of streets and lots in the new community.

When a surveyor makes a survey of a lot the first thing he does is to establish a *point of beginning* which may be a mark on a walk or a corner stake. This point serves to help him locate the other corners of the piece of ground and also to relate levels (elevations) to the city requirements for that particular lot.

The surveyor shows much other information on his plat besides the location of the lot corners. He also indicates the location of utilities, and water and sewer mains. Unless the lot is almost flat he shows the contour of the lot by using a series of contour lines taken at a uniform interval of levels or grades. He shows trees, walks, and streets on his drawing. See Fig. 9-2.

A surveyor's plat is usually required by banks and other lending institutions before mortgages for building can be arranged.

Scales Used by the Surveyor. A surveyor uses a tape or chain, and a leveling rod graduated in feet and tenths of a foot. He makes his drawing using a scale of 1" = 10' or 1" = 20'.

A dimension of 120.5' would mean 120

and 5/10 feet or 120 feet and 6 inches. A dimension of 65.7' would mean 65 and 7/10 feet. In order to convert 7/10 of a foot to inches the following process is used:

$$\frac{7}{10} \times 12'' = \frac{84''}{10}$$

$\frac{84''}{10} = 8.4$ inches or almost 8½ inches. For practical purposes in laying out a lot or a building, the dimension used by a craftsman would be 65'-8½". (To be more accurate, the Decimal Equivalent Table in the appendix indicates that .4 inches is closer to ⅜ of an inch.)

Fig. 9-2 shows the elevation of the corners of the lot as 112, 108, 111.2 and 114.9 feet. The latter two elevations would be approximately 111'-2½" and 114'-10¾".

Elevations. The height of any point on the lot is called its *elevation* and is measured above or below some point of reference. (The use of the word elevation in this sense must not be confused with the elevation of a house which is a view of the outside of the structure.) In Fig. 9-2 the elevation of the point of beginning is 112 feet. This means that the level of this point is 112 feet above the point of reference used in that community. It may be above sea level, above mean lake level if there is a large body of water nearby, or it may be above some fixed point on a street or building. The elevation of all other points on the lot are measured by using a transit and a leveling rod in reference to the datum point. See Fig. 9-3.

Contour Lines. Contour lines are lines drawn on a surveyor's plat or on an architect's plot plan to show the slope of the ground. A contour line may be explained by thinking of the lowest contour line as the shore of a lake. If the water were to rise one foot the shore line (contour line) would take a new shape as the water covered more of the earth. The intervals of contour lines on small lots are usually in units of one foot elevations. If the tracts are large or very hilly, contour lines may be in two foot or larger units.

The Plot Plan. The architect uses the information supplied by the surveyor's plat in drawing the plot plan. He draws the lot accurately, usually converting the dimensions from feet and decimals of a foot to feet and inches. He indicates the north point, the lot and block number, the datum point, or point of beginning taken from a local city measuring point, and locates utilities, sewers, streets, and public walks. He then locates the house, making sure that he has observed all of the area restrictions, front and side limitations, etc., established by local ordinances. He draws the contour lines showing the *natural grade* elevations. He then draws a second set of con-

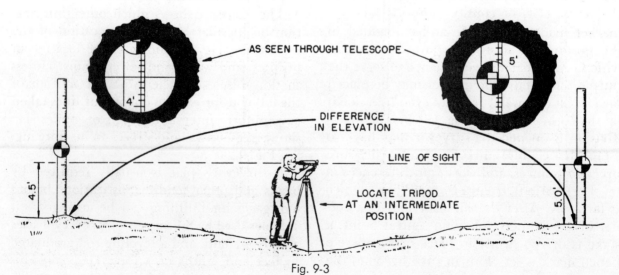

Fig. 9-3

Information on elevations is determined by using a transit and leveling rod.

The Plot Plan 107

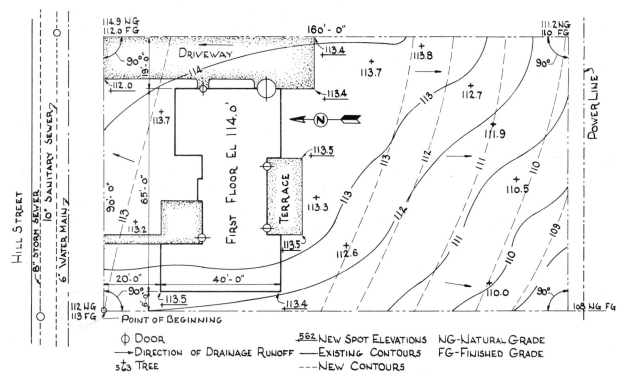

Fig. 9-4

An architect's plot plan shows elevations, location of the house, and much other information.

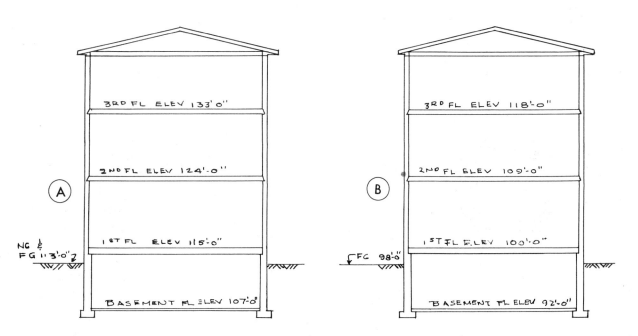

Fig. 9-5

Elevation of floors may be derived from datum levels as at A. The elevation of the first floor may be set at 100'-0" for convenience as at B.

tour lines to show how the lot is to be sloped after the project is completed. These lines will determine the *finish grade*. See Fig. 9-4.

The level of the point of beginning given by the surveyor is designated, and the first floor is indicated so the excavator may know how deep to dig the hole for the foundation.

In the plot plan, Fig. 9-4, the architect established the first floor level at 114'-0" in direct reference to the datum point. When a plan may be used more than once on different lots, the elevation of the first floor is often shown as ±0'-0". (Note that this is the manner of designating the floor levels in Chapter 11.)

When tall buildings are built each floor is designated by an elevation. The architect may use this city datum information as a base as shown in Fig. 9-5 A. The first floor level is 2'-0" above the natural grade which makes it 115'-0". The basement floor level is 6'-0" below the natural grade which makes it 107'-0".

The architect has the option to establish the level of the first floor at 100'-0" in order to make measurements for other floor levels easier to determine. See Fig. 9-5B. The first floor must first be established in proper relation to the natural grade. Then the level of the finished grade is changed to the correct relationship to the first floor level.

Self-Check Quiz No. 9

(Based on Fig. 9-4)

Complete the following statements. *Answers are given in the appendix.*

1. The dimensions of the lot are ____ × ____.

2. The corners are each ____ degrees.

3. The long dimension of the lot is in a _____ compass direction.

4. The greatest dimensions of the house are _____ × _____.

5. The house is located _____ from the West property line.

6. It is located _____ from the North property line.

7. It is located _____ from the South property line.

8. The natural grades at the four corners are _____, _____, _____ and _____.

9. The greatest difference in the natural grade elevations of all points on the lot is _____.

10. Most of the water would run off of the lot toward the _____ corner.

11. The first floor elevation is _____.

12. The terrace elevation is _____.

13. The driveway slopes a difference of _____ from south to north.

14. The area on which the house stands is to be graded to 113′. The first floor elevation is _____ above the finished grade.

15. List the utilities, water, and sewer mains shown: _____
_____

TEST DRAWING NO. 8 (See pages 111-112)

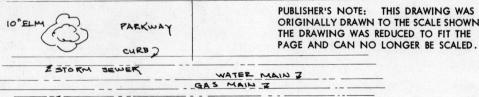

Trade Competency Test No. 8

(Chapter 9: Based on Test Drawing No. 8)

STUDENT'S NAME _____   INSTRUCTOR'S NAME _____

COMPLETION TEST: Key words or specific information has been omitted from each of the following statements. Fill in the spaces.

For Scoring

1. The north and south lot lines (are or are not) _____ parallel.

    1._____

2. All references to elevations refer back to the _____ shown on this plot plan.

    2._____

3. The natural slope of the ground is downward toward the _____ and _____ corners of the lot.

    3._____

4. The existing front walk has a difference of elevation of ____ foot from west to east lot corners.

    4._____

5. The lowest point of the lot in its original condition is _____ feet elevation.

    5._____

6. The highest point of the lot in its original condition is _____ feet elevation.

    6._____

7. After grading is finished, the grade around the house will be _____ feet elevation.

    7._____

8. After grading the amount of slope toward the point of beginning corner will be _____ (foot, feet).

    8._____

9. After grading the amount of slope toward the NE corner will be _____ feet.

    9._____

10. Water, gas, and sewer mains are found in the _____.

    10._____

11. Electric power is supplied from a _____.

    11._____

12. The elevation of the finished first floor is _____ feet.

    12._____

13. The change of elevation from the finished first floor to the garage floor is _____.

    13._____

14. The garage floor is not level. The slope is _____ foot.

    14._____

15. The apron drops a maximum of _____ foot from the garage opening to the walk.

    15._____

(Cont.)

Trade Competency Test No. 8 *(Cont.)*

16. The dimension from the house to the front property line is _____.

16._____

17. The dimension from the house (omitting the chimney) to the west property line is _____.

17._____

18. The dimension from the house (not including the garage) to the east property line is _____.

18._____

19. Dimensions are all shown in feet and _____.

19._____

20. Elevations are all shown in feet and _____ of a foot.

20._____

SCORE: ☐

Chapter **10**

Reading Blueprints for Trade Information

Everyone concerned with the construction of a building should be able to read blueprints. The owner will want to know about every feature of his future home. Real estate brokers, bankers and loan agents will want to know how to appraise it also. However, the greatest burden for the accurate reading of the blueprints rests with the contractor, sub-contractors and all of the craftsmen. They must know how each part of the structure is made, where its equipment is to be located, and what problems they might encounter in their own fields.

Each craftsman must be able to read the parts of the blueprints which pertain to him with perfect understanding and also be able to read the prints in an overall sense so that he can work intelligently with men from other trades. He must be able to sift out those things which do apply to him and understand those which do not apply. Cooperation between craftsmen is essential because the work of one man hinges on that of another in many instances. The carpenter sets the window and door frames for the brick mason. He must also provide floor and wall framing for the plumber before he installs the piping and fixtures for the bathroom. Sheet metal workers must provide flashing and gutters in cooperation with roofers, carpenters and brick masons. A fine sense of good will is generally found among workmen as they work toward the common objective of creating a building in the most efficient manner.

The blueprints, particularly the plan views, have their limitations because they would be hopelessly cluttered if all of the details of construction were shown. The craftsman learns how to interpret sectional views, see Fig. 7-2, 8-5, and details such as Fig. 8-6 to help him understand the overall structure. However the blueprints are basic, he must see far beyond the information shown on them. The carpenter must picture the whole structure in his mind as shown in Fig. 7-1, and then cut and fasten each member in place. The electrician must run wiring, the plumber run piping and the sheet metal worker install ducts in both horizontal and vertical directions between the various structural members. The competent craftsman can solve many of the problems before construction begins and make provisions for them. He solves other problems as they arise during the course of construction.

On large construction jobs the architect and his staff do the original planning and prepare what is known as the architectural working drawings; these are then printed as blueprints. A separate set of working drawings is prepared by a structural engineer covering the details of the steel or reinforced concrete skeleton. Another set of working drawings is called the electrical set which goes into detail regarding electrical equipment and circuits. A fourth set of working drawings covers the mechanical work which includes heating, air conditioning and plumbing. In a small structure the architect's drawings must suffice. The heating contractor will usually make his own drawing to show where he intends to run ducts, the electrical contractor will sketch the various circuits on a set of blueprints, and the plumbing contractor will make a drawing showing how the piping will be arranged and the pipe sizes to be used.

Types of Trade Information on Blueprints. The ability to read blueprints on the part of the contractors and craftsmen centers in the following.

1. Trades are generally required to make a cost estimate of the work to be done on the building covering a certain aspect such as excavating, painting, electrical work, etc. The blueprints must be analyzed for quantity studies so that the bid will be realistic and competitive.

2. The details of construction to be followed by the workmen are indicated on plan, section and detail views.

3. The craftsman must use the blueprints in "taking off" material in the actual phases of construction. He can achieve many economies in the use of material by planning ahead on such items as structural members, sheathing, wall covering, trim, piping, and conduit.

4. The plans indicate the precise location of building features such as windows, doors, mechanical equipment, and fixtures.

5. Tradesmen study the blueprints to solve problems requiring the cooperation of different trades.

Division of Labor. There may be as many as fifteen different kinds of craftsmen who work on a building in the course of construction. There may be some overlapping where one man is qualified to do more than one job and has the skills required. Also a particular type of work may be done under the jurisdiction of one trade group in one part of the country and another trade group in another part of the country. One contractor may hire men of more than one trade so that he can present one bid covering more than one trade area.

For instance, roofing and sheet metal might be included in the same contract so that men of different trades will cooperate on the actual work.

Specifications. A set of specifications is included in the work of the architect. It is a set of typed sheets ranging from a few pages for a simple house, to a book with hundreds of pages for a large building. It supplements the blueprints and has equal value. It takes up in detail the general information of the legal responsibilities of the contract, guarantees of performance, permits, supervision of construction, etc. Then each area of work is described in detail, specifying the materials to be used and how they are to be installed. (The subject of specifications is discussed in *Building Trades Blueprint Reading, Part 2.*)

The topics for the divisions in a set of specifications also indicate, to a certain extent, a breakdown into trade areas. A typical specification for a small home would include the following:

1. General Requirements
2. Excavating—Filling and Grading
3. Concrete
4. Masonry
5. Carpentry
6. Roofing and Sheet Metal
7. Plaster and Drywall
8. Special Interior Covering
9. Glass and Glazing
10. Painting and Decorating
11. Heating and Air Conditioning
12. Plumbing
13. Electrical
14. Landscaping
15. Hardware (when required)
16. Metals, Structural and Misc.

Modular and Component Concepts. The use of modular units and component parts in buildings has steadily increased in volume and has brought about new concepts in several areas of the industry where they may apply. The craftsman has moved indoors into factories and uses some of the assembly line techniques developed in the automobile industry. In the modular system, whole rooms are built as units such as complete bathrooms which are delivered to the job and dropped into place by huge cranes. The module is, in a sense, a building block in the form of a whole room with fixtures installed ready to connect to other parts of the building. Some modules consist of half a house cut longitudinally. Each part is completely finished. The units are narrow enough to be transported on trailers. They are slid into place on the foundation and fastened together along a long center wall. Plumbing and heating are then hooked up. The building is then ready for occupancy within a few days.

Components are built in a shop and are delivered to the job as complete wall sections with windows as well as exterior and interior finish applied. Other components are floor and roof sections. The unit parts for both systems are assembled on jigs which are holding devices designed so that the members are held in precise relation to each other before they are fastened together.

The craftsmen who lay out the modules or components and set up the jigs must read the blueprints accurately and make precise measurements. The parts must fit together perfectly. The work done on the site in the building of the foundation and installing water supply and drainage piping to hook up to the building must be done in advance and be done with great care if the building is to fit in place.

Modular dimensioning and planning is a different concept than the use of modules mentioned above. Modular dimensioning is a system of design in which all measurements of the building and its parts, both horizontal and vertical, fall on modular lines which are made by a grid with 4 inch squares. Most manufacturers of material are doing everything possible to make their products fit this 4 inch grid pattern. See Fig. 10-1. Generally the work of the architect is made more difficult because he must work out every detail

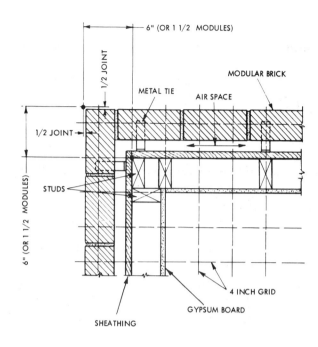

Fig. 10-1
A detail of a corner of a brick veneer building shows how masonry and wood units are placed in relation to 4 inch modular grid lines.

to fit the grid. Craftsmen must work with greater precision because they must make units of brick, stone, sheathing, flooring and wall covering fall at exact intervals.

The eventual result, if modular planning is to be used on a wide scale, would be the opportunity to use standard components for building parts and the reduction in cutting and fitting of material.

Electrical Work

The electrician has several guidelines to follow which tell him how to proceed with the work of providing an efficient electrical installation which has no shock or fire hazard. First of all he must have a fundamental knowledge of electrical theory so that he may know the facts about electrical energy and how it is controlled in circuits. He also should have an extensive knowledge of the many kinds of conductors and wiring devices as well as other electrical apparatus he is to install. His general background should include a working knowledge of the National Electrical Code which is a set of standards covering the design and manufacture of electrical devices and materials and the manner of their installation. Fire insurance policies are generally written on the basis that the building is built to follow this code. He must also be familiar with the local electrical code for his city or state. This usually includes the National Electrical Code provisions and then goes on to cover local restrictions.

The electrician must then be familiar with the part of the specifications for the particular building which deals with electricity. This specification which is drawn up by the architect includes a statement of the work to be done and then describes the materials and fixtures to be used giving exact descriptions of each, often including exact details such as catalog numbers and trade names. Some of the specific items include the size wire to be used and the type of wiring, whether nonmetallic sheathed cable, armored cable, rigid conduit or thin wall tubing. See Fig. 10-2. Lighting fixtures are usually chosen by the owner but are purchased through the contractor. The architect specifies a certain sum of money in the specification for this purpose. The electrical contractor includes this amount in the bid for the job and then pays the amount after the owner chooses the fixtures.

The architect's blueprints (see Fig. 5-8 for a typical plan view) show where outlets are to be located in the finished structure and how switches operate to control them. In addition to the architect's blueprints, an electrical blueprint or wiring diagram is provided or made by the electrical contractor. See Fig. 10-3. In a small building this may be a small drawing which shows how the circuits are arranged. In a large building or in a building with complex electrical needs, a set of blueprints are prepared by an electrical engineer showing the wiring in great detail. The electrician draws on his knowledge of installation practices and the information supplied by the wiring diagram or electrical blueprints when he proceeds to his work.

The blueprint reading part of his work requires that he study the blueprints of the house to know the general layout of the building and its structural features. It will make a difference to him if the building has balloon or platform framing, if the joists run in one direction or the other, if the attic space is accessible for wiring, or if there is a crawl space or basement. His main concern, however, is the location of outlets, their purpose and how they are connected to switches and special equipment. These might be fans, bathroom heaters, water heaters, clothes dryers or other equipment requiring special wiring provisions. See Fig. 5-8. A table of electrical symbols (Fig. 5-4) should be memorized so that he makes no mistake in reading the blueprint.

Type NM nonmetallic sheathed cable comes with a ground conductor. Conductors are insulated with a color coded polyvinyl chloride (PVC) compound. The cable is covered with a tough moisture and abrasion resistant PVC jacket.

Type AC armored cable has insulated conductors and a flexible galvanized steel strip cover.

Rigid conduit is non-flexible steel tubing threaded at both ends.

Electrical Metallic Tubing (E.M.T.) is known as "thin wall." It is lightweight tubing and does not require threaded ends.

Fig. 10-2
Basic types of wiring. The tubing and conduit are shown without insulated conductors.

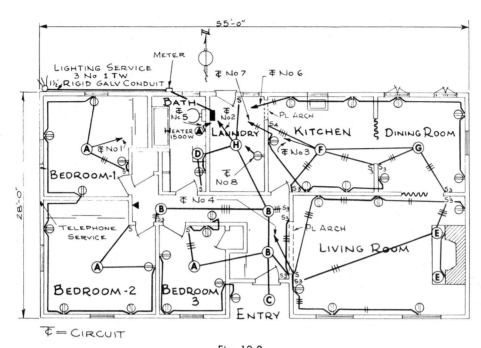

Fig. 10-3
A wiring diagram tells the electrician how the circuits are to be arranged.

Basically, no dimensions are given to show how far outlets are to be placed from room corners or doorways. The exact location is usually left to the electrician who must work within the limitations of the building structure and the provisions of the National Electrical Code. A few inches one way or another will have little bearing on the usefulness of most outlets. A broken line connects the outlets and switches. See Fig. 5-8 and floor plans on pages 60, 73 and 74. These lines indicate how outlets are connected and where the switches are that control them. The lines are only for this purpose and do not show where the wires are run. The electrician has the job of making these connections in the walls, under floors and over ceilings in as short a path as he can devise. The circuits are shown on the electrical blueprint or the wiring diagram.

The electrical blueprint or wiring diagram, see Fig. 10-3, shows the separate circuits in the building which are protected by individual fuses or circuit breakers in the distribution panel. Fig. 10-3 shows the service drop from the power line coming into the Northwest corner of the building and carried in conduit to the meter and then to the distribution panel. Wherever there is equipment such as an electric range, water heater or other heavy duty appliances a 240 volt service requiring three wires is necessary. The conductor specified is No. 1 TW (thermoplastic) wire. The circuits which are independent of each other and which terminate in the distribution panel are designated on the plan by number. Circuit 1 for instance includes all of the outlets for bedrooms 1 and 2, see Fig. 10-4. The arrow in Fig. 10-3 shows that the line returns to the distribution panel. Small cross lines on the lines connecting outlets indicate the number of conductors in each conduit. The branch circuit schedule (Fig. 10-4) will help in following each of the eight separate circuits. Circuit number 5 serves only one outlet which is a 220 volt bathroom heater. The letters on the ceiling outlets, Fig. 10-3, refer to the Lighting Fixture Schedule, Fig. 10-5, and help in determining which fixtures are to be installed in respective places and also the number which are alike. It will be noted that 3-way switches

BRANCH CIRCUIT SCHEDULE							
CIRCUIT	LOCATION	LIGHTING OUTLETS		SWITCH OUTLETS			RECEPTACLE OUTLETS	SPECIAL
		CEILING	BRACKET	SP	3-W	4-W	G	
1	Bedroom-1	1		1			4	
	Bedroom-2	1		1			4	
2	Bathroom		1	1			1	
	Laundry	1		1				
	Bedroom-3	1		1			4	
	Hall and Entry	4			4		2	
3	Kitchen	1			2	1		
	Dining Room	1			2			
4	Living Room		2	1	2		6	
5	Bathroom							1
6	Kitchen						2	
	Dining Room						3	
7	Kitchen						3	
	Dining Room						1	
8	Laundry						1	

Fig. 10-4
The branch circuit schedule gives details of each circuit.
SP—single pole
3-W—3 way
4-W—4 way

LIGHTING FIXTURE SCHEDULE								
CIRCUIT	TYPE							
	A	B	C	D	E	F	G	H
1	2							
2	1	3	1	1				1
3							1	1
4					2			

Fig. 10-5
The lighting fixture schedule gives information regarding the location and type of fixtures.

are used in the living room and also in the hall. Three-way switches are necessary to make the proper connections when two switches at different locations in the room may be used to turn the same outlets on and off. The lights in the living room can be turned on or off either near the front hall arch or near the dining room opening. Also to be noted is that one of the switches in the kitchen is a 4-way switch. This is required whenever the same outlets can be turned on or off at more than two places. The kitchen light is controlled by switches near the opening to the laundry, at the door to the hall and near the dining room opening.

The circuits are arranged by the architects, the contractor or an electrical engineer so that the requirements of the National and local codes are observed. Although each circuit is protected by a fuse or circuit breaker, overloading must be avoided. When overloading does occur there is a loss of power to lights and appliances and energy is lost in the form of heat. Also when lines are very long; a voltage drop occurs which lowers the efficiency of lights, motors and appliances. The size of the wire is an important factor in preventing overloading. The circuits must be designed so that the wires can carry the designated amount of current without power loss. The National Electric Code states that one lighting circuit should be provided for each 500 square feet of house area calculated from the

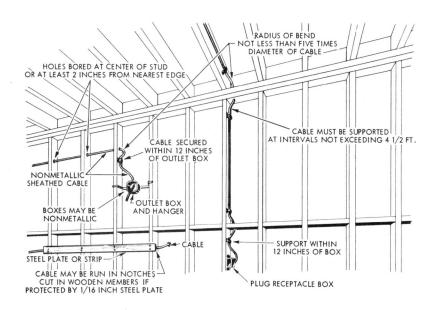

Fig. 10-6
Nonmetallic sheathed cable is installed with staples or through holes in structural members.

outside of walls. A 20 amp appliance circuit requires a No. 12 size wire and a 15 amp circuit requires a No. 14 wire. The area regulation for lighting is minimum. More lighting may be necessary thus requiring additional circuits.

With all of the technical information in mind and the electric wiring diagram in hand the electrician is ready to install the wiring in the building under construction. He uses one of four basic types of wiring depending on local codes.

1. Nonmetallic Sheathed Cable. See Fig. 10-2. In nonmetallic sheathed cable, two insulated conductors are enclosed in a tough cover of polyvinyl chloride compound. Most local codes require that a third bare copper wire be included in the cable so that it can be grounded. It is desirable that a continuous ground be provided back to the distribution panel. This is done by fastening the grounding wire to a ground terminal on the receptacle, motor or other electric device and to the metal outlet or junction box. The cable itself is fastened to the wood framing members of the building with staples at specified intervals. Junction and outlet boxes are equipped with clamping devices to hold the cable firmly at the end of each run. A study of Fig. 10-6 will show some of the restrictions on wiring with sheathed cable in an actual installation. Compared to the other systems it is relatively easy to install.

2. Armored Cable. Metal clad armored cable is a fabricated assembly of insulated conductors in a flexible metallic enclosure, see Fig. 10-2. It is frequently called "B-X". It too contains a continuous wire or metal strip called a bond wire which is fastened to the enclosures or receptacles at each end of a run.

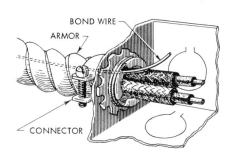

Fig. 10-7
Metal clad armored cable is clamped firmly at each outlet box. The bond wire provides a continuous ground.

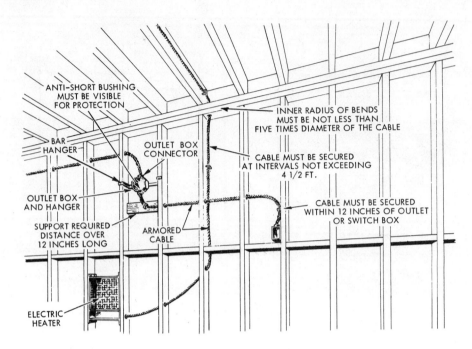

Fig. 10-8
A typical armored cable installation shows its versatility.

The armor is cut with a hack saw or suitable tool at the required length and an anti-short bushing is inserted which protects the insulation on the wires. The cable is fastened to a connector which is then inserted into the outlet box and fastened securely with a locknut. See Fig. 10-7. Armored cable is easy to install because of its flexibility. Fig. 10-8 shows how practical it is because it can be threaded through studs and along structural members with ease.

3. Rigid Conduit. Rigid conduit (see Fig. 10-2) is required for wiring by the National Electrical code and by local codes for specific applications. It comes in trade sizes from ½ inch to 6 inches (nominal inside diameter). The smaller sizes are bent by hand using a hickey. See Fig. 10-9. A hydraulic bender is required for the larger sizes. The conduit comes in 10 foot lengths and is threaded at both ends. Whenever it is cut it must be reamed out so that no sharp edges might cut the insulation on the conductors. It must be rethreaded to provide for threaded couplings and for locknuts and bushings at boxes. The conduit is installed during the course of construction of the building when the studs and joists are exposed. The wires are pulled

Fig. 10-9
Conduit is bent using a hickey.

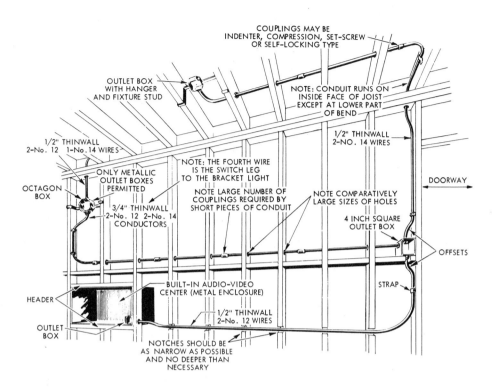

Fig. 10-10
Electrical metallic tubing (thinwall) is commonly used in residential work.
Note: Notching of studs is preferred over drilling holes for tubing.

through later usually after the wall finish is in place.

4. Electrical Metallic Tubing. Electrical metallic tubing (EMT) known as "thinwall" is only about 40% as thick as rigid conduit and does not need to be threaded. See Fig. 10-2. The ½ and ¾ inch thinwall is generally used on residential work but the tubing can be obtained as large as 4 inches in diameter. Actually, ¾ inch tubing is 0.824 inches ID (inside diameter) and 0.922 OD (outside diameter). Thinwall tubing has replaced rigid conduit to a great extent in small home wiring because it is bent with greater ease and has couplings and box connectors which do not require threading. Fig. 10-10 shows a typical installation of thinwall steel tubing, indicating how studs and joists are notched or drilled and how the tubing is bent to enter the boxes.

The electrician picks up the wiring of the house at the power drop outside of the house. He provides the wiring to the meter usually through rigid conduit and then makes connections to the distribution panel. Depending on the type of wiring required, he proceeds to run wire, armored cable, or conduit to meet all of the electrical needs of the house. His knowledge of general construction will help him run his lines with a minimum of waste and effort.

Plumbing

It is easy to take modern plumbing for granted. The bathroom is expected to be a showplace in new homes with beautiful fixtures placed in a setting of elegance. The kitchen too, a place where plumbing plays an important part, is also very attractive. The plumbing system is one of the least understood installations in the whole house yet perhaps the most important convenience. What occurs within the walls and under the floors to provide a supply of sparkling clean water

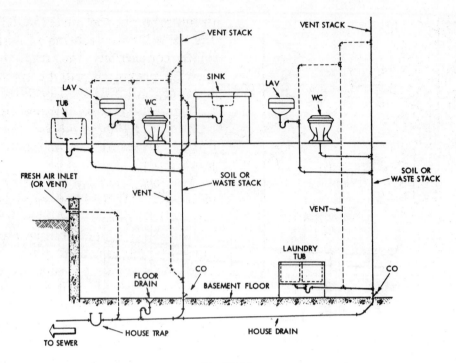

Fig. 10-11
This figure illustrates a typical sewage disposal and venting system for a residential dwelling.

and also to carry away the waste water and sewage is only the plumber's concern as long as these systems provide perpetual trouble free service. The plumber has considerable responsibility to plan and execute his work to provide this service. There is a great hazard to health if the piping is not designed correctly. In most cities and states plumbers take examinations and are certified or licensed before they can work at their trade in the full capacity of journeyman plumber.

The plumber has the need for reading the blueprints of the building the same as other tradesmen. His first interest of course is the location of all of the plumbing fixtures in bathrooms, kitchen, and laundry, as well as the location of water supply lines and sewer lines. He studies the structure of the building to see if he will have special problems in running his pipes in the walls. He is particularly concerned to find out how piping shall reach upper floors through the lower floor partitions. He will want to know the direction of joists in the floors in order to know how the horizontal piping will run and what problems he will have under the bathroom floors.

The building ordinances in some cities require that the architect prepare a plumbing drawing to show the plumber how to hook up the drainage system. This may be in elevation form (see Fig. 10-11) or in isometric (see Fig. 10-12). Generally, however, the plumber or contractor must make his own plumbing diagram which represents a slice through the building taken to show the fixtures and piping.

After studying the blueprints of the house, the plumber will read the specifications for plumbing written by the architect to apply to the particular building. The specifications generally contain, (1) a statement of the scope of the work, (2) a list of materials to be used for water, drain, and gas pipe, (3) information on installation, (4) a list of the fixtures giving complete descriptions including trade names and sizes and (5) information on the type of tests for the water system and the soil system.

The plumber must be familiar with the

Reading Blueprints for Trade Information

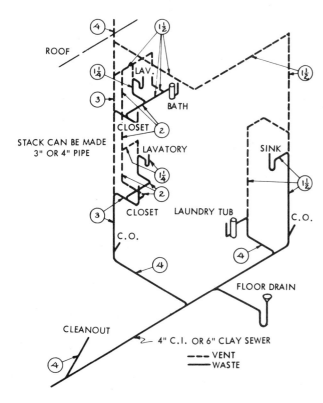

Fig. 10-12
The plumbing layout for a two story house shows the waste system and the manner of venting. Numbers in circles indicate size of pipe "in inches."

National Plumbing Code and the local code which supplements it. The codes cover items such as the kind and size pipe which is permitted for each use covering water supply, drainage, vents, and buried underground piping. It will cover venting provisions, fixture waste connections and trap requirements. Testing provisions will be spelled out. For instance: Soil pipe shall be filled with water and tested for leaks. Water pipes shall be tested under 100 pounds per square inch hydrostatic pressure for one hour and leaky joints replaced.

Materials. There are five basic kinds of pipe used today: (1) steel and wrought iron pipe, (2) cast iron pipe, (3) seamless brass and copper pipe, (4) copper tubing and (5) plastic pipe. In addition, fiber pipe, vitrified clay pipe, and concrete tile pipe are used for the outdoor parts of drainage systems, particularly in connection with septic tank fields. The common types of pipe joints are: (1) screwed (threaded), (2) bell and spigot, (3) flanged, (4) soldered, and (5) welded. Each of these have their own symbols. If the symbols for the screwed type are learned, the other types may be easily understood. For example, study the symbols for a 90° elbow shown in Fig. 10-13. Note the similarity of the symbols for the different types of joints. The plumber must be able to cut and assemble the pipe using the methods and fittings appropriate to each type. Besides pipe and fittings there are a great number of other devices which he should know about, such as valves, traps and other plumbing parts.

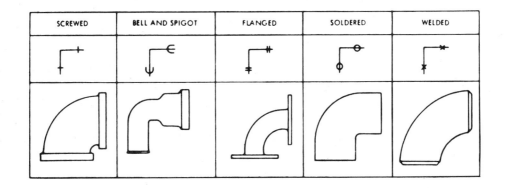

Fig. 10-13
These five basic types of 90° elbow pipe fittings are shown with the name and joint symbol above each fitting.

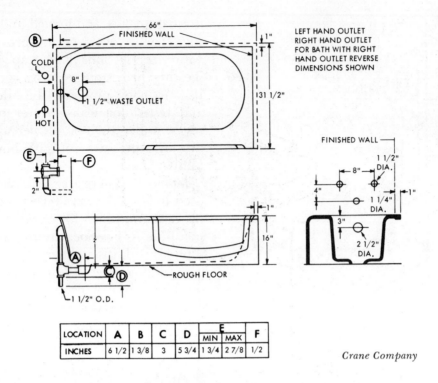

Fig. 10-14
The roughing-in diagram shows the exact location of piping.

Procedure. Some of the work of the plumber is done very early in the construction sequence of building such as when he makes connections with the street sewer, runs the house drain which will be under the basement floor, and brings the water supply into the house. However, most of his work will be done after the house is roughed in with all of the walls and partitions in place, but with studs exposed and floor joists accessible. At this time he installs all of the supply, waste and vent piping in walls and floors leaving stubs which will protrude at the proper place after the walls are plastered or covered with drywall. He will need the help of the carpenter who will cut and reinforce the floor joists under bathrooms to provide support for the soil pipe and heavy fixtures. The sinks and lavatories are installed after the wall finish is in place. Most of the new fixtures are supplied with faucets which are surface mounted on the fixture instead of on the wall thus making it easier to place hot and cold water lines. The tub which is installed when the building is roughed in is an exception. Fig. 10-14 shows a typical roughing-in sheet supplied by the manufacturer showing how the tub is to be supported and the locations of the pipe outlets for waste, overflow and faucets.

Water Supply. Water enters the house from a city water main under pressure or from a well through an individual pump. See Fig. 10-15. A stop and waste valve located at the lowest point in the system is placed in the line to permit the system to be shut off and drained. Cold water lines are run as directly as possible to each faucet in bathrooms, kitchen, laundry and to hose bibbs outside the house. A branch line feeds the hot water tank and a main hot water supply line leaves the tank and divides into branches to go to each fixture requiring hot water. Shutoff valves are supplied at fixtures wherever there is a likelihood for the need of repair. Air chambers which are only continuations of the sup-

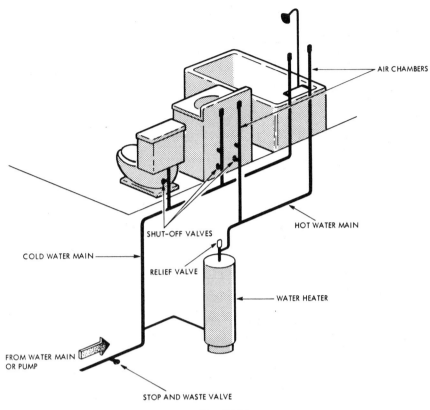

Fig. 10-15
The water supply lines deliver clean water to fixtures.

ply pipe above the faucets serve to prevent the noise known as water hammer when the water is shut off suddenly. See Fig. 10-15. It is important that there be no cross connections with waste water lines.

Drainage System. The disposal of waste and waste water is the most important part of plumbing, requiring great care to prevent contaminated water from returning into the supply lines. A companion pipe system to the supply lines is required, but completely isolated from them. A system of soil and waste stacks of specified size are worked out to carry water away from each fixture. Those carrying waste from toilets are called main or soil stacks. See Fig. 10-11. The water is carried downward with a slight pitch to the vertical soil or waste stack. These stacks carry the water to the house drain which in turn leads to the sewer away from the building. The main stacks are usually 4 inches and if cast iron pipe is used the partition in which the pipe is run must be 6 inches or more in thickness to accommodate the wide hubs of the pipe. The soil or waste stacks project above the roof as vents.

A water seal is required at each fixture. This is done by means of P shaped traps and a drum trap at the tub. The house drain under the basement floor is pitched at least ¼ inch per foot and equipped with a U shaped trap which serves to keep sewer gas from entering the house. Cleanout fittings are placed at the bottom of each stack and at any turns in the house drain so that the plumber can open the drain to rod it out if blockage occurs.

Sometimes the house drain cannot be located below the basement floor because the street sewer is above this level. The house drain may have to be hung from the ceiling or run along a basement wall. A sump pit and sump pump are installed in the basement floor to collect water from basement floor drains, fixtures and appliances such as the laundry tub and washing machine. Wherever water collects to a certain level the sump pump begins to operate raising the water to enter the house drain.

Venting. Venting the system serves the purpose of equalizing pressure within the system. If there were no vents the water seal in the traps might be lost and the flow in the drainage system retarded. See Fig. 10-11. Venting also serves the purpose of removing gas and odors. The vent extends parallel to each soil or waste stack from a low point to join branch vents from each fixture. The vent joins the soil or waste stack above the highest fixture branch which then continues as a vent stack through the roof. The vent stack projects at least 8 inches above the roof. A fresh air inlet may be required in the house drain under the basement floor just inside the U trap to serve as an escape for air and gases and to let air into the system.

Sewage Disposal. Sewage disposal is a problem for the plumber in areas where there is no public sewer which leads to some form of sewage disposal plant. In rural areas sewage from individual houses is routed into a septic tank and then into an absorption field where it is dissipated in the ground. The septic tank depends on bacterial action to break down the waste matter. Decomposed solids remain in the tank and the liquid overflow flows into the absorption field. See Figs. 10-16 and 10-17.

Storm Drains and Drain Tile. A problem for home builders in many parts of the country is how to divert large volumes of rain water. Usually the local code does not permit this water to enter the sewer unless it is very large or there is a separate storm sewer system which is adequate to take it. In many locations rain water is diverted from the gutters and downspouts and run to a place some distance from the house where it is permitted to run out on the ground. In other communities it is directed to dry wells where it is permitted to seep out gradually to be absorbed by the earth. Surface water near the building often presents another problem. The plumber lays drain tile around the house at the level of the footing. It is laid with open joints in beds of gravel to receive the water which seeps down near the foundation. The water may be diverted to a dry well or, if the local code permits, the drain tile may be connected to the sewer.

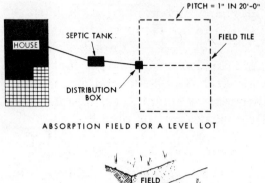

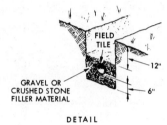

Fig. 10-17
The liquid from the septic tank is distributed through field tile to be absorbed by the earth.

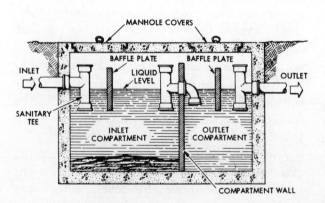

Fig. 10-16
The septic tank reduces sewage by chemical action of bacteria.

Gas Piping. Gas piping is brought into the house and to the meter by the gas company. The plumber installs piping from that point to each fixture requiring gas for fuel. Copper or steel is used depending on the code restrictions. Generally gas piping presents little difficulty for the plumber.

Sheet Metal Work

Sheet metal workers are skilled mechanics who work in several areas of the building trades. In home construction they protect exposed places on the exterior of the building from water damage by the use of flashing; construct and install gutters and leaders and build sheet metal canopies and decks. Inside of the building they are concerned with installing heating and air conditioning plants. This subject will be covered in the next section *Heating and Air Conditioning*. In commercial and industrial buildings there are many things for the sheet metal worker to do. He makes and installs such things as conveyors, complex duct systems for heating, air conditioning and dust collection, machinery guards, and hoods. One division of the trade is devoted to the manufacture of cabinets, lockers, and shelving.

Many of the fittings and parts are stock items which he simply assembles. However, he is required frequently to fabricate parts in the shop to meet special requirements and must be able to modify them on the job to fit the irregularities in the structure. He must have a high degree of skill in laying out the patterns and in cutting, bending, shaping and assembling the finished products. See Fig. 10-18. He must be familiar with the charac-

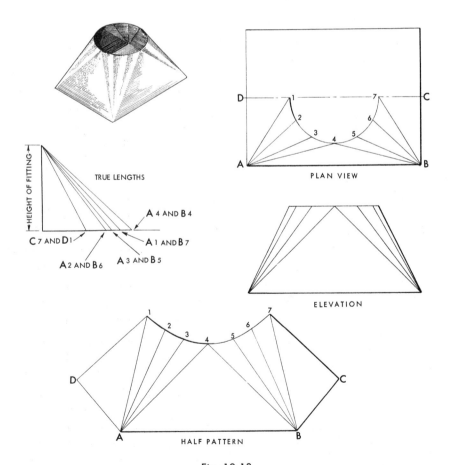

Fig. 10-18
The sheet metal worker learns how to lay out fittings such as this square to round transition piece.

teristics of his materials and know their limitations as he bends shapes and fastens them together. Although most of the time he works with galvanized sheet steel, other metals including copper, aluminum, lead, zinc, tin, stainless steel and alloys such as monel metal may be used. Plastics are entering the market and used for gutters, leaders (downspouts) and other applications. His general skills include the ability to make fittings using several types of seams fastening them by soldering, riveting or welding. There are a great number of bolts, nails, brackets, clamps, and hangers which should be part of his stock in trade.

Reading the blueprints for exterior sheet metal items is not difficult because they are generally pointed out on elevation or sectional views. The specification written by the architect on this subject amplifies the drawings with instructions on the type of metal to be used and the manner in which it is to be applied.

One of the first uses of sheet metal in the construction of a building in areas where termites are found is the termite shield which is placed around the top of the concrete foundation wall. See Fig. 10-19. It is bent to shape and made continuous by overlapping joints around the perimeter of the wall. Flashing is one of the more general applications. The elevations of the house shown on pages 27 and 28 indicate flashing where the

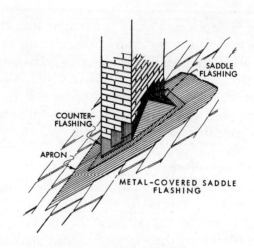

Fig. 10-20
Flashing at the chimney provides a waterproof seal.

garage roof intersects the house, over the bay, and at the chimneys. Wall flashing is sheet metal in the form of strips 8 feet long and 12 inches or more in width. It is bent down the center into a right angle and nailed along the edges. The ends are overlapped at least 2 inches. It is nailed to the sheathing of the wall and of the roof to make a waterproof covering of the intersection of the two building parts. Flashing at the chimney can become quite involved. See Fig. 10-20. A sheet metal saddle is designed to shed the water from behind the chimney. Step flashing is used on the sides of the chimney itself with each piece turned into brick joints. Counter-flashing which is a second thickness of sheet metal placed over the other sheets in a reverse fashion to cover the edges finishes the job. The roofer lays the shingles carefully so that the water does not find its way under the edges of the flashing.

The flashing around a dormer is made watertight and weathertight by means of overlapping metal shingles and strip flashing at all edges which intersect the roof. See Fig. 10-21. Flashing is required over heads of exterior doorways and other exterior millwork which presents flat surfaces where rain or snow might collect. It is also used over the water table (at the bottom of the siding) of

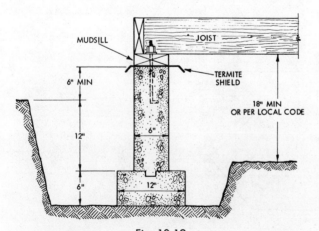

Fig. 10-19
A termite shield is used between the concrete foundation and wood sill.

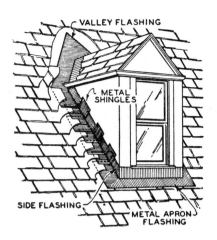

Fig. 10-21
Sheet metal flashing is used at all edges where a dormer intersects a roof.

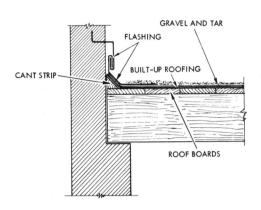

Fig. 10-22
Flashing is used at the intersection of a brick wall and a flat roof.

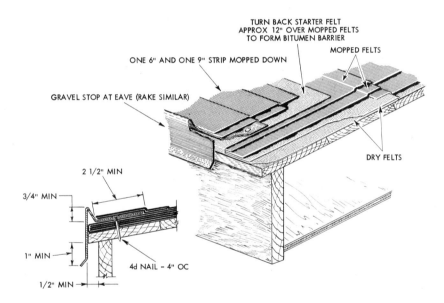

Fig. 10-23
Detail shows gravel stop and manner of placing felt strips.

frame buildings for certain wall coverings such as shingles. The intersection of the roof and a parapet wall on a brick building is protected by flashing also. See Fig. 10-22.

Sheet metal plays an important part in applying the roofing to the building and carrying away rain water. An edge strip of metal is used under the first row of shingles. When the roof is flat or nearly flat a built up roof of several layers of felt mopped with tar is used and the edge strip takes the form of a gravel stop. See Fig. 10-23. Valleys are often lined with a strip of sheet metal bent to fit snugly against both sides of the valley. It may be crimped to keep the water from rushing down one slope to cross over to go under the shingles on the other slope. See Fig. 10-24. Sheet metal gutters are placed at the edge of the roof over the sheathing and the face of the fascia (see Test Drawing No. 6, page 86,

130 Building Trades Blueprint Reading

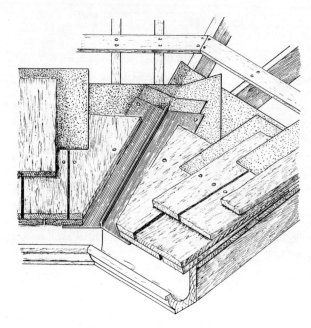

Fig. 10-24
A sheet metal valley is crimped to provide a water barrier on a roof with hand split shakes.

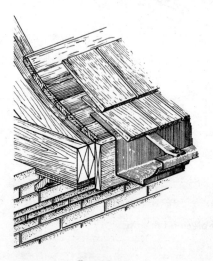

Fig. 10-25
A sheet metal gutter is attached to the edge of a roof.

sect. A-A) before shingles are fastened in place. See Fig. 10-25. Straps are used to support the gutter at 3 foot intervals. The gutters and leaders which bring the water to the ground come prefabricated in standard shapes with corners and other fittings. Very often the gutters are made for the particular job in the shop or at the job site. The sheet metal worker solders the connections so that the gutters are continuous. The gutters are sloped to provide drainage to the leaders.

Canopies over bays or extended entrance doorways, metal roof decks, roof ventilators and other special exterior sheet metal work require skillful layout and accurate forming skills, as well as careful installation so that they may resist water and wind damage indefinitely.

Heat and Air Conditioning

Heating and air conditioning require the skill and "knowhow" of several craftsmen. Forced air heating, cooling air plants and air conditioning require ducts and are installed by sheet metal workers. Hot water and steam systems are installed by pipe fitters. Electric heat is installed by electricians who also bring the wiring to furnaces and boilers in other systems. Carpenters are often called on to help provide openings in walls and floors to accommodate ducts and pipes. Gas lines are provided by plumbers.

The craftsmen who are responsible for installing the heating plant should study the blueprints to find out where the furnace or boiler is to be located and where the radiators, diffusers or convectors are to be installed. They should know the direction of the floor joists and the arrangement of studs in the walls and partitions so that they can plan horizontal and vertical runs of ducts or piping.

The specifications will help to give them information about the heating unit itself, the controls, and pertinent information on how the system is to be installed. The architect rarely gives them a detailed layout of ducts or pipes. This is a problem for the heating contractor. The contractor usually has an engineer make a layout of the ductwork or piping in detail with information as to the size of ducts or pipes, radiators, convectors and diffusers required to deliver the proper amount of heat to each room. The sizes are determined by careful calculation of heat

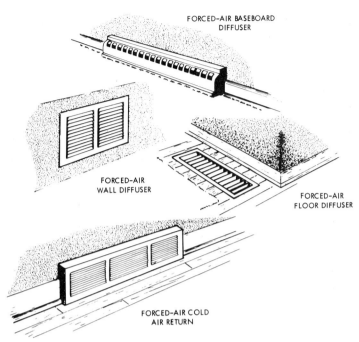

Fig. 10-26
Wall diffusers and returns provide circulation in a forced air system.

loss (and heat gain) in each room because of window area and other factors. The problem of installing the furnace or boiler are not too great. The difficult part is usually the running of ducts and pipes following the engineer's layout and meeting the problems presented by the structural makeup of the building.

Heating Systems. There are four fundamental types of heating used in homes: (1) forced warm air, (2) hot water, (3) radiant panel heat, and (4) electric heat. The type of fuel whether coal, gas or oil used in heating a forced warm air or hot water system is important only in that it provides a source of heat. Each of course has its own characteristics which affect the design of the heating unit.

Forced Warm Air Heat. Forced warm air is popular as a means of heating a home because it is efficient, quick to respond and can be readily adapted to air conditioning. The main elements are: (1) a furnace in which fuel is burned, containing, (2) a heat exchanger which is an enclosed space where the air is heated, (3) a duct system designed to bring the right quantity of warm air to diffusers in each room, and (4) a means of returning cold air to a cold air chamber in the furnace. A high velocity fan is installed in the cold air return at the furnace to keep the air circulating rapidly throughout the system. The air is changed four to six times per hour in an average five or six room house. The temperature of the air is about 155°F when it leaves the furnace arriving at the rooms at about 140°F. Warm air outlets are generally placed on outside walls and under windows to counter the down draft which occurs at the windows. Cold air returns are generally placed on inside walls. Each room except the bathroom and kitchen are usually supplied with a cold air return. See Fig. 10-26. Small homes often have only one or two centrally located cold air returns. Several blueprint symbols commonly used for forced air heating systems are shown in Fig. 10-27.

Two types of warm air heating systems are

132　Building Trades Blueprint Reading

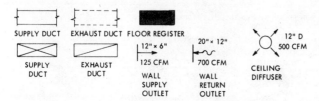

Fig. 10-27
Blueprint symbols for forced warm air systems.

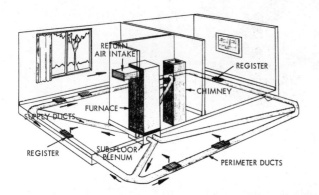

Fig. 10-28
A warm air perimeter loop system has a flow of warm air to registers at the outside walls and a single return intake.

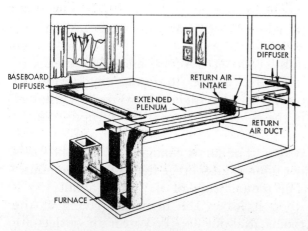

Fig. 10-29
A forced warm air extended plenum system has individual ducts projecting to baseboard diffusers.

shown in Fig. 10-28 and Fig. 10-29. In the perimeter loop system the duct is continuous around the building with one central cold air return. In the extended plenum system a large trunk which is an extension of the furnace warm air plenum, carries warm air across the basement ceiling. (A plenum is an enclosed space where air is kept under pressure and connected to ducts.) Individual ducts fan out to each room and are controlled by dampers which serve to balance the system. The adjustment of the dampers permits more or less air to go to individual rooms as desired. (A sheet metal worker who is an expert in heating should do the balancing.) The main control of the furnace is the thermostat which is located in one of the rooms where it will operate best to call for heat or shut off heat to maintain the comfort level. There are other controls on the furnace which turn the blower on and off for the efficient use of fuel. Each furnace is equipped with a humidifier which supplies moisture to the system as the air passes through it and electronic air cleaners may be obtained.

The sheet metal worker installs the furnace which may be delivered to the job in parts. He lays out the ductwork according to the engineer's sketch adapting it to the structure of the house. In a brick building provisions have been made by masons who leave channels in the brickwork to provide for the ducts to rise to the second floor where the carpenters provide similar openings in the partitions. In frame buildings duct spaces are made in exterior walls also. The ductwork is put in before wall finish is put on the walls. Registers and deffusers are put in place after the wall finish is in place but before the base trim is put on by the carpenter. The duct work may be quite complex. See Fig. 10-30. Most of the parts are stock items, but the sheet metal worker is required frequently to cut and make adjustments and also to make fittings in the shop to care for special problems.

Central Air Conditioning. The forced warm air system is readily adapted to central air conditioning. A cooling coil is placed in the plenum of the furnace so that the blower will force air over it and on through the house in the summer time. In the winter the warm air passes through the coil which is not in opera-

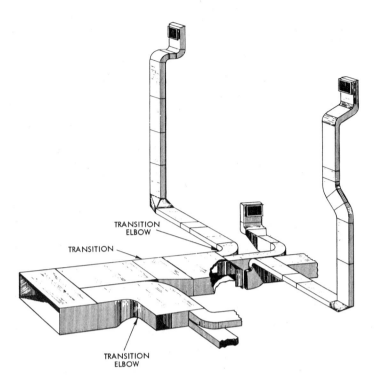

Fig. 10-30
Duct work may be very complex. The sheet metal worker must visualize the system.

tion. The other important parts of the air conditioning system are the condenser which is usually placed outside of the house, and the controls which are designed to control both warm air and cool air. Fig. 10-31 shows a schematic sectional view of a year-round air conditioning system. The return air is circulated through a cooling coil which is then recirculated through the building. The complex controls include thermostats, humidity controls, outside air control and an outside thermostat so that the system can detect changes in the outside temperature.

Hot Water Heat. Hot water heat under pressure called *hydronic* heat is practical and popular for use in homes. Steam heat and gravity hot water systems have been replaced by this system. An automatically fired gas or oil burner heats the water in the boiler which is distributed to radiators or baseboard convector units by means of a circulating pump. See Fig. 10-32. A one-pipe forced hot water system is popular in small homes because it is easy to install and produces even heat in all of the rooms. See Fig. 10-33. The single pipe follows the perimeter of the house with branches and risers connecting to the individual convectors. When the thermostat calls for heat, the flow control valves open and the pump starts. Special fittings allow restricted amounts of water to enter each connector so that the heat going to each convector varies only slightly. A two-pipe system provides separate piping for the hot water and for the return to the boiler. Water expands when heated. The expansion tank permits the water to expand and helps keep the system full. The pipe fitter has the problem of cutting and fitting pipe between the various units of the system. He must understand the function of each part of the controls and piping and install them correctly.

Radiant Heat. Radiant heat through the use of radiant heat panels in the floor or ceiling is an effective way of heating a home. This is especially true of homes with slab-on-ground foundations. Instead of surrounding the occupants of the house with circulating

134 Building Trades Blueprint Reading

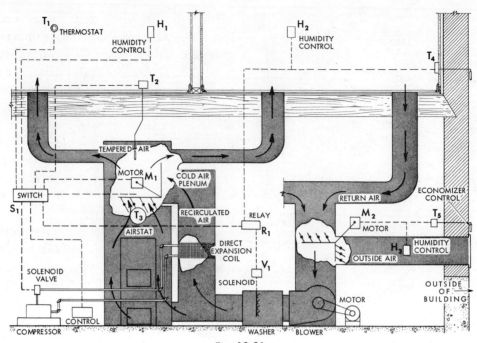

Fig. 10-31
A typical control system for a year-round air-conditioning system. A mechanical furnace is the basic heat source.

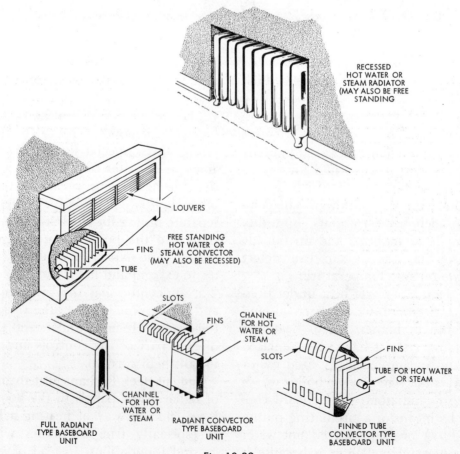

Fig. 10-32
Convectors serve to transfer the heat from the hot water to the air.

Reading Blueprints for Trade Information 135

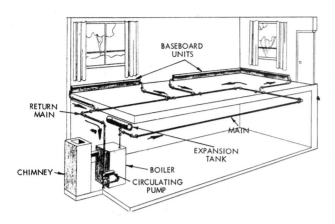

Fig. 10-33
A single pipe forced hot water system provides efficient heat.

warm air they are placed in an environment in which the objects in the room as well as the floor, ceiling and walls serve as reflectors of radiated heat. When radiant panels are located in the floor a layer of concrete is laid first. The pipe fitter then places a serpentine coil of tubing or a field of connected piping in place to be covered by another bed of concrete. A boiler and pump are installed with controls similar to the hydronic system. See Fig. 10-34. Radiant panels may be applied to ceilings instead of floors with the piping embedded in the plaster. This type of heat calls for the cooperation of cement finishers, plasterers or drywall applicators, because it demands care so that the tubing or piping not be damaged.

Electric Heat. Electric heat has been used to heat individual rooms with individual resistance units for many years. In climates without severe weather this has been extended to serve the heating needs of the whole house. Today great advances in devising heating systems, in bringing down the cost of electric energy, and more efficient means of insulating have made electric heat more popular in colder climates. Baseboard units which serve

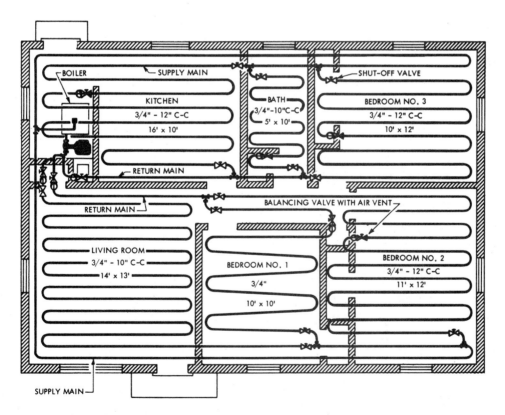

Fig. 10-34
This is a typical arrangement for a radiant heating floor panel for a five-room residence.

136 Building Trades Blueprint Reading

to heat by both radiation and convection are operated by individual thermostats or a zone thermostat.

Some units operate by means of an electrical heating element immersed in water. The heated water circulates into tubes which remain warm for a considerable time after the thermostat shuts off the electricity, thus extending the heat cycle.

Concrete Work and Masonry

Concrete work for light construction is divided into the following: (1) erecting forms and pouring foundation walls, and (2) the laying of concrete walks, driveways, basements and garage floors. Carpenters generally erect forms for foundation walls and cement finishers lay walks, driveways, and floors. Craftsmen who work with concrete must study the blueprints carefully so that foundations are exactly right in size, height, thickness, placement and driveways laid correctly for proper drainage. Note that Chapter 9, The Plot Plan, should be reviewed regarding the location of points on the lot in relation to the *point of beginning* and regarding estab-

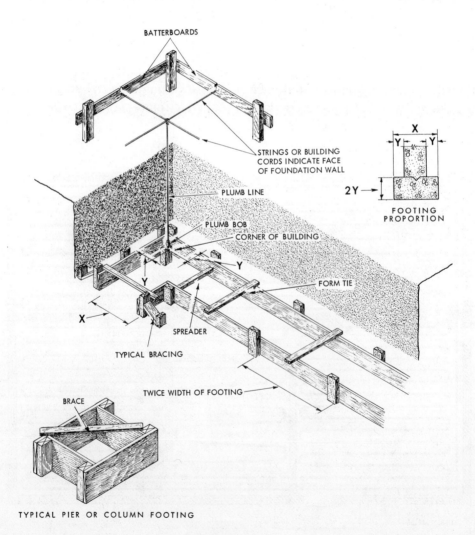

Fig. 10-35
A point is dropped from the intersection of the building lines and placed on a stake. Footings and the face of the foundation are located from this point.

lishing elevations. See Fig. 9-3. The plot plan on page 110 and the basement plan on page 73 should be studied also. They are from the set of blueprints which include pages 27, 28, 60 and 74.

Concrete Foundation Work. Concrete foundation work begins with the establishment of the corners of the building in relation to the *point of beginning* for measurements. Stakes are set up which give the excavator the information he needs for the size and depth of the excavation. The next step is to erect batterboards from which strings are stretched to indicate the outside face of the foundation wall, see Fig. 10-35. The batterboards are set back from the excavation and are used for checking purposes at several steps in the construction of the footings and the erection of foundation forms. Footings are usually poured first as a separate operation. The formwork for footings is placed so that the foundation wall will be centered on the footing and so that the top of the footing will be level and at the proper height. Once the footing is poured and has had a chance to set, the formwork for the foundation walls may proceed. The carpenter studies the plot plan again, page 110, to check the dimensions from the property lines and to note the finished first floor elevation. He studies the section views, see page 86, for details of the foundation wall itself, and the elevation views, pages 27 and 28, for information on special forming problems. He uses the basement plan, page 73, for details of the actual shape of the foundation.

Forming System. There are several forming systems in use, depending on the scale of operation and the preference of the contractor or carpenter. Some formwork is assembled using sheets of plywood and wales. Wales are 2 x 4 inch horizontal members used to hold the forms in line and to provide stiffness. Clamps and ties hold the plywood sheets in position. See Fig. 10-36. Other patented formwork may be very elaborate, designed for easy assembly and many repeated uses. One of the more common types of formwork uses 4 x 8

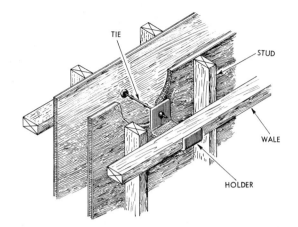

Fig. 10-36
Patented form ties and holders are used to hold wales against studs.

foot carpenter made panels, 2 x 4 inch wales, and snap ties which hold the formwork together. See Figs. 10-37 and 10-38. The carpenter often makes a layout of the formwork so that he need not waste time on the job figuring out each step. Fig. 10-39 is such an analysis of the foundation forming for the basic house (See page 73.) using 4 x 8 foot forms, with 2 x 8 foot special forms and fillers to make up the required dimensions. The step footings at the garage require special forms to be cut and fitted at the job. (These stepped footings are required in order to make a transition on firm earth from the house to the garage.) The outside forms are erected first, beginning at one corner. They require bracing in order to stand and are made plumb (vertical) and are anchored before the inside forms are placed. Holes are made opposite each other in the inside and outside forms, the ties inserted, the clamps put in place, and metal wedges, which are part of the clamp, are tapped with a hammer to bring the whole assembly up tight. See Fig. 10-38. The wales serve to line up the forms, add some stiffness and to provide surfaces for the clamps to bear on. The top of the concrete in the form is marked by nails which are set by means of the carpenter's level or level-transit so that the top of the wall will be

138 Building Trades Blueprint Reading

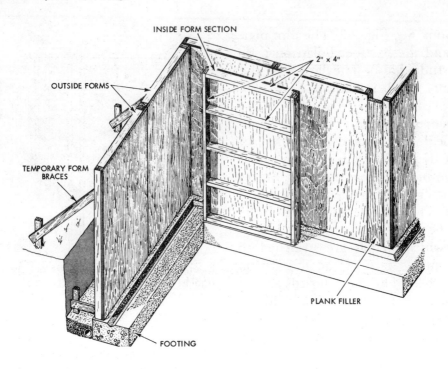

Fig. 10-37
Panels with fillers of various sizes can be adapted to most forming problems. They may be reused many times. Forms of this type are generally made by the carpenter.

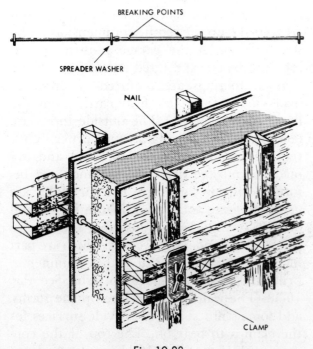

Fig. 10-38
Snap ties provide a means of spacing the walls at the proper distance apart, and also clamping the whole assembly together.

correct in accordance with the blueprints. The foundation wall is then poured and the forms are left in place for several days until the concrete has set. After stripping, the ties are broken off inside the wall (see Fig. 10-38) and the rough place in the wall repaired with grout.

Slab-on-ground foundations in northern climates require footings and foundations which extend below the frostline. When frost leaves the ground in the spring, the earth heaves with considerable force. Footings and foundation walls, unless they extend below the frost line, will tend to raise and crack. A crack will be a means for water to seep through the wall. The frost line varies in different localities depending on the sustained low temperature. Edge insulation is required and footings for interior bearing walls may take the form of a continuous beam. See Fig. 10-40. In warmer climates the forming is quite simple. A perimeter support may be

Reading Blueprints for Trade Information

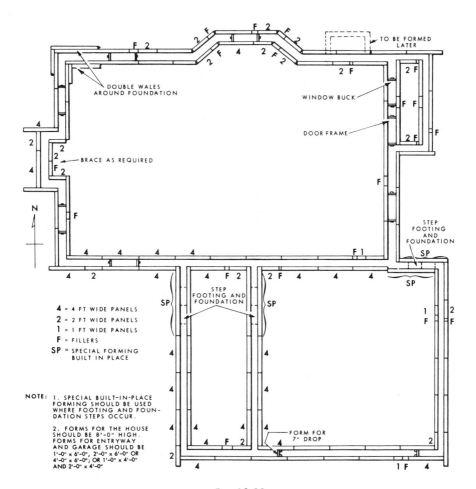

Fig. 10-39
A panel layout helps the carpenter save time as he erects the forms.

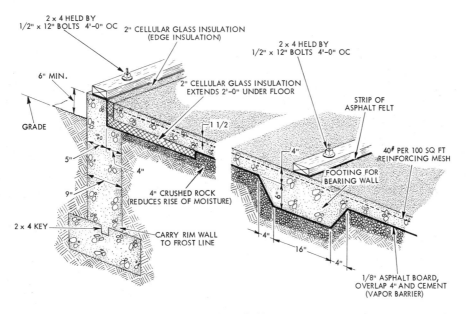

Fig. 10-40
A concrete floor laid on the ground requires a vapor barrier, and in cold climates must have edge insulation.

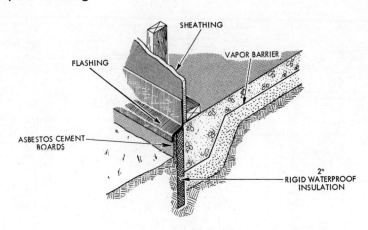

Fig. 10-41
A simple perimeter support will serve for light construction in warm climates.

sufficient (see Fig. 10-41) or a shallow foundation wall may be made of concrete block.

Laying Walks, Drives and Floors. The walks, drives and garage floor will be laid at some time late in the building program when the earth has had a chance to become compact and little activity is anticipated at the site so that the concrete can have time to set and can cure without being disturbed. The floor may be laid in the basement at the same time, provided the plumber has laid the house drains which will be under the floor. The cement finisher will study the plot plan supplied by the architect (see page 110) for dimensions and elevations at various points. He will erect forms using boards of sufficient width to retain the concrete and will support them with stakes and braces. He should be careful to check the dimensions and should use his builder's level or level-transit to see that the surfaces will come to the right elevation and slope to drain. The concrete is usually delivered as ready-mix and spread over the area. Driveways and garage floors are often reinforced with a heavy steel mesh or a grid of reinforcing rods to give them strength to resist cracking. After the pouring, the cement finisher goes to work skillfully working the concrete to a smooth surface with the desired pitch. The surface should be kept wet for several days as the concrete cures and must not be driven on until it develops strength which would be several days more. Forms are removed when convenient after the second day.

Masonry. Masonry includes the laying of brick, concrete block, stone and tile. (See pages 83 and 84 for brick veneer construction and masonry buildings.) The work is done by brick layers or stone masons, men who have developed the skill of laying the units according to plan with precision and speed. The responsibility of the bricklayer or mason demands that he study and make frequent references to the blueprints. Perhaps he studies the elevation drawings first to find out where brick or stone is used. The sectional views through exterior walls will tell him how the masonry units are to be laid in the wall, at windows and doors, and how the top of the wall is to be finished at the eaves. They will also show the type of construction whether solid brick known as ordinary construction, brick or stone veneer, or concrete block. See Figs. 7-10, 7-11, 7-12 and 7-14 for isometric views of brick applications. The most important blueprints are the floor plans which give the exact location of windows and other openings. The bricklayer must read the plans accurately in order to place the window and door sills in the proper places. The carpenter sets the frames for the bricklayer and

Reading Blueprints for Trade Information

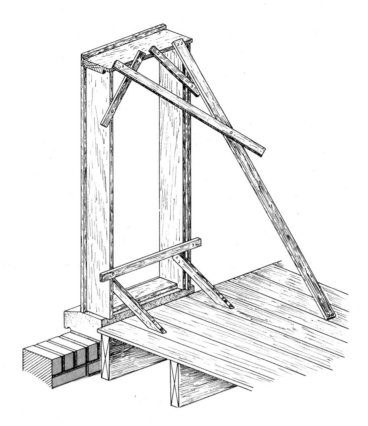

Fig. 10-42
The carpenter sets the frames in a brick wall to make sure that they are plumb and located according to plan.

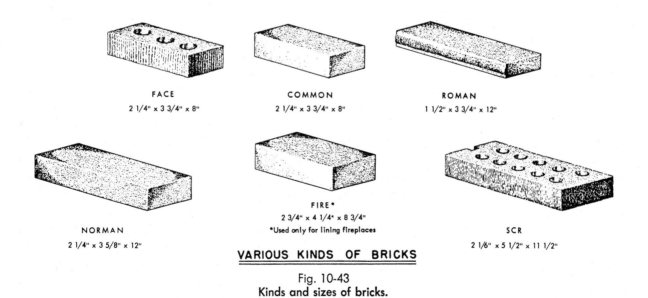

VARIOUS KINDS OF BRICKS

Fig. 10-43
Kinds and sizes of bricks.

checks their location. See Fig. 10-42. The bricklayer lays the rows of brick so that they are perfectly level and plumb, and at the same time arranges the bricks so that they fill the space lengthwise with little or no cutting. He is required to lay the courses with

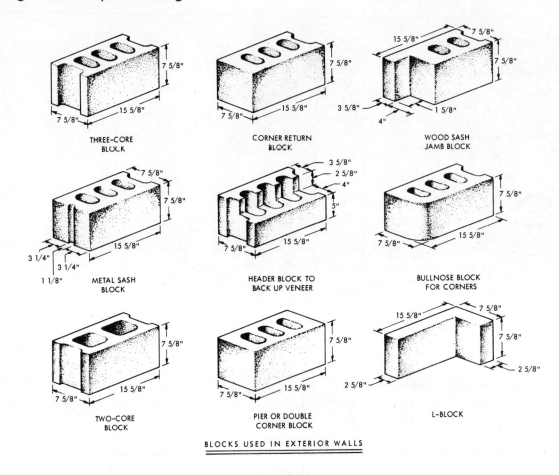

Fig. 10-44
Concrete blocks used in exterior walls.

care vertically maintaining the same mortar joint so that lintels above openings fall in place at the right level and the level from one floor to the next floor be the required dimension and the exact number of courses. Typical brick and concrete block sizes are shown in Fig. 10-43 and Fig. 10-44.

There is an increased emphasis on the use of modular bricks which come in several nominal sizes depending on the thickness of the mortar joint. One typical size modular brick to be used with a ¼ inch mortar joint, is actually 3¾ x 2⅜ x 7¾ inches. The bricks are laid with great care and with accurate joints in order to make the wall surfaces work out to areas based on a 4 inch module. See Fig. 10-1.

Brick, stone, and special cement block are used for decorative purposes as well as for strong walls. Brick may be laid in several different bonds (see Fig. 10-45) and in unusual patterns. The laying of stone is often developed to a fine art. Each stone is chosen to fit into the wall to give a pleasing effect. Cement block (textured blocks) can be obtained with decorative faces or pierced designs to give interest to exterior and interior masonry walls.

Welding

There are no persons classified as welders on the construction job, but many craftsmen find welding to be an essential skill spending much or all of their time joining metal by this process. Others find that welding is desirable if not essential. Structural iron workers join steel members by means of welding in the course of every day's work. Pipe fitters and plumbers

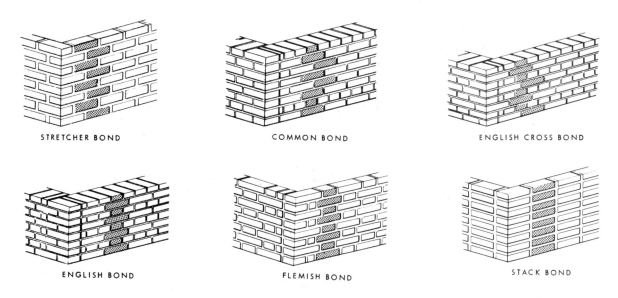

Fig. 10-45
Brick may be arranged in a variety of bonds.

join large pipe and pipe which must withstand heavy internal pressure using this method. Some carpenters, sheet metal workers, and electricians find it essential to be able to weld when they are involved in particular operations.

Structural iron workers use riveting, bolting and welding to fabricate the structural parts in the shop and to assemble them on the job. Welding is used exclusively on some jobs. When the main columns, girders, and beams are assembled by bolting, welding is used for secondary structural parts such as supporting members for masonry, for intermediate supporting members, and for fastening steel decks. The flicker of the electric arc is a part of the building of a steel structure. The structural iron worker accomplishes his job under adverse conditions of height and often poor accessibility to the place to be welded, yet he must make a satisfactory weld which will safely carry the load for which it is intended. Blueprint reading definitely fits into his program. The structural engineer prepares a set of drawings covering the basic design of the structure which are referred to the steel fabricator. His draftsmen will make a shop drawing for each piece, giving dimensions, location of rivet holes, and a description of each weld. A shop drawing will be made to show details of sub-assemblies.

An assembly drawing is often made to show how the pieces are to be fastened on the job indicating the type of welds to be used. The structural iron worker must understand how to read the detail and assembly drawings and decipher the welding symbols. See Fig. 10-46.

Pipe fitters and plumbers find that they are required to weld steel pipe which must be continuous without fittings to withstand high water or steam pressure, gas, oil, or chemicals. Their work must be perfect. An imperfect weld could easily cost lives. Furthermore they very often find it necessary to work under the adverse conditions of great height, below ground level in a trench, or among a maze of pipes over head. To learn to weld around a pipe in whatever position it fits into the piping takes years of practice.

Carpenters' building methods are being adapted using metal studs and light metal runners. Their ability to weld becomes more valuable constantly as they install metal building parts.

Sheet metal workers often use welding to

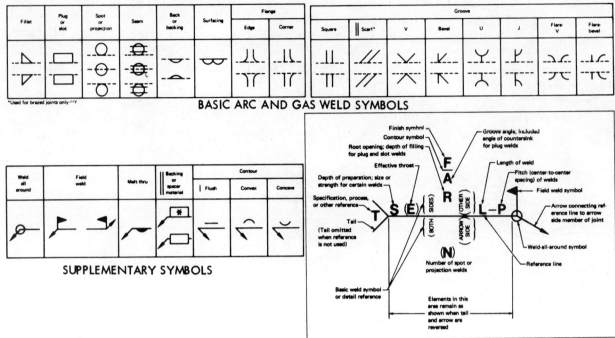

Fig. 10-46
Basic welding symbols.

American Welding Society

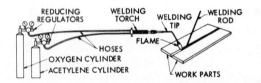

Fig. 10-47 *General Motors Corporation*
Most commercial metals can be welded successfully by the oxyacetylene process.

fabricate sheet metal products and to fasten sheet metal parts to the steel structure of the building. They must develop special skills because they are working with thin metal and with galvanized metal which loses its zinc coating when heated to the temperature of the welding process. Special welding techniques have been developed to solve their problems.

Electricians also find welding skills valuable. Among the jobs requiring welding are the construction of hangers and the fastening of electrical equipment to steel structures.

There are two main processes used to weld metal (1) Oxy-acetylene welding and (2) electric arc welding. Oxy-acetylene welding, Fig. 10-47, is done by the combustion of oxygen and acetylene at the point of a torch. Directing the flame to the edges of the metal to be joined, the intense heat causes a molten pool to form. A metal filler rod held in position near the molten pool, becomes molten and adds to the metal. By changing the tip of the torch, the oxygen supply can be increased so that the equipment can be used to cut metal.

In arc welding, an electric arc formed between the work and the electrode liberates enough heat to effect fusion at the point of the weld. See Fig. 10-48. The intense heat brings a small portion of the metal to the melting point. The electric arc is formed across a gap between an electrode held by the craftsman in a special holder and the work itself. The tip of the flux coated electrode is melted and the tiny globules of metal are added to the molten pool. Electric arc welding is used almost exclusively on construction jobs whenever the power is available.

Reading Blueprints for Trade Information   **145**

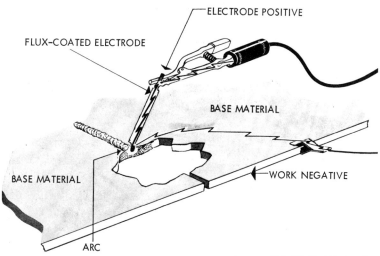

Fig. 10-48   *Republic Steel Corporation*
Arc welding is the most generally used type of welding. The electric circuit is completed through the electrode and the base material with the arc jumping the gap between the two.

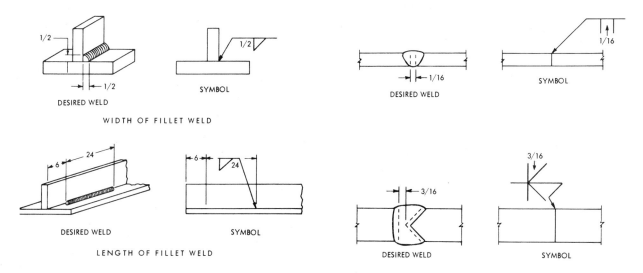

Fig. 10-49
Symbols are used to indicate the size, and type and location of the weld.

There is a certain amount of danger involved in all types of welding: highly flammable gases, heavy electric current, the danger of burns, injury to eyes, flying molten slag and metal globules and smoke and gases resulting from combustion are all hazards to be dealt with. Proper training, good equipment, correct clothing, shielding devices, and much experience in safe welding practice minimize the dangers involved.

For most craftsmen who weld, the ability to read blueprints is applied in the fabrication of flat stock and structural steel shapes and fastening them as part of the building structure. The pipe trades craftsmen have the added responsibility of working with the welding of pipe. The significant part of reading the drawings is to understand the application of the welding symbols. See Fig. 10-46. Four sample welds and their symbols are shown in Fig. 10-49.

Self-Check Quiz No. 10
(Based on the text material in Chapter 10)

Complete the following statements. *Answers are given in the Appendix.*

ELECTRICAL WORK

1. The basic rules for the industry are contained in the _____.
2. A wiring diagram for a residence shows how the outlets are arranged into individual _____.
3. Most non-metallic sheathed cables contain a base copper wire to provide a _____.
4. Armored cable is useful because it has the quality to be _____ and good protection for the conductors.
5. Electrical metallic tubing is easier to install than rigid conduit because _____ is not required for couplings or box connections.

PLUMBING

6. Water supply lines are kept completely isolated from _____ lines.
7. Waste and soil stacks lead to the _____ under the basement floor.
8. Gas is prevented from entering the house because of a _____ at each fixture.
9. One of the main purposes of providing vent pipes parallel to the soil and waste stacks is to _____.
10. The septic tank is used to hold sewage so that it may be broken down by _____.

SHEET METAL

11. Exterior sheet metal items are shown mainly on the _____ views.
12. Some of the methods used to fasten seams are riveting, soldering and _____.
13. Metal pieces used to protect places on the exterior of a building where water might enter are called _____.
14. Finishing flashing which overlaps other flashing is called _____.
15. The sheet metal rain trough at the edge of the roof is called a _____.

HEATING AND AIR CONDITIONING

16. Warm air diffusers are usually placed on _____ walls.
17. The important parts of an air conditioning system are the duct systems, the cooling coils, the fan, the controls and the _____.
18. Uniform temperature in a house using forced hot water heat is maintained by means of a thermostat, a circulating pump and several _____.
19. Radiant heat is a type of forced hot water heat with coils of pipe in the _____, _____, or walls.
20. Electric heat uses _____ units.

CONCRETE WORK AND MASONRY

21. The location of footings, foundations, walls and drives is determined in relation to the _____ on the lot.
22. The members used to line up a concrete formation are called _____.
23. The cement finisher uses a builders level or level transit to see that the concrete surfaces have the correct _____ and _____.
24. Carpenters and bricklayers are responsible for the correct placement of frames for _____ and _____.
25. The manner of laying brick causes to create different patterns and effects on the face of the wall is called _____.

WELDING

26. Structural iron workers fasten steel structural members together with bolts, rivets, or _____.
27. The steel detailer will prepare a series of _____ drawings showing how each piece is to be cut and welded.
28. Pipe fitters weld pipe to withstand _____ steam.
29. The oxy-acetylene method of welding uses a _____ to add to the metal in the molten pool.
30. In the arc method of welding the circuit creating the arc gap is completed through the _____ and the _____.

TEST DRAWING NO. 9A (See pages 153-156)

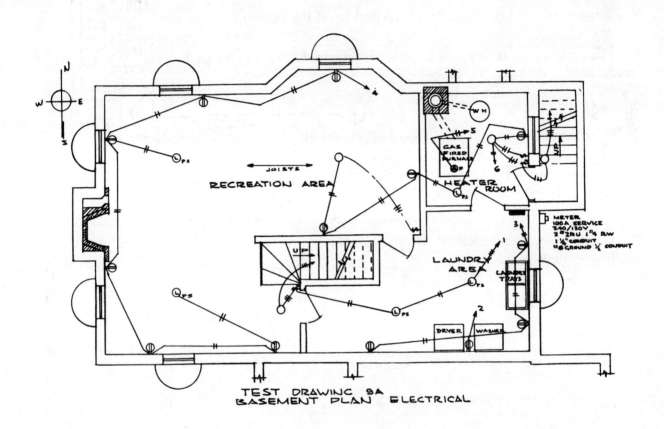

Reading Blueprints for Trade Information **149**

TEST DRAWING NO. 9B (See pages 153-156)

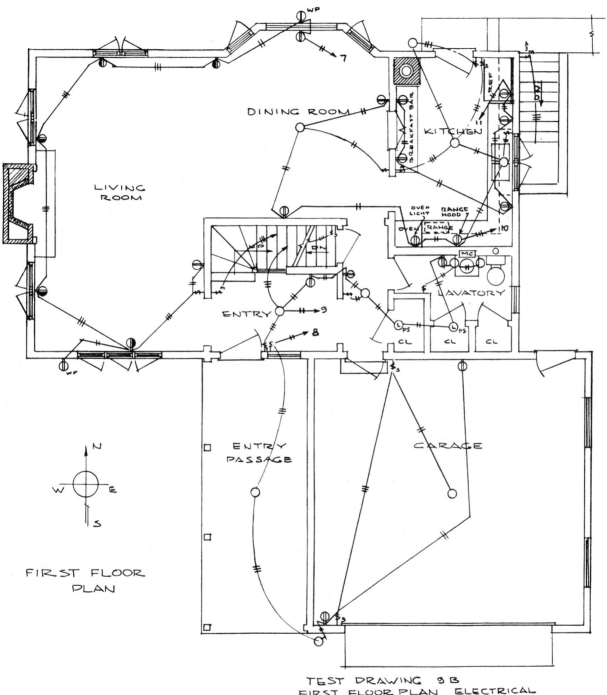

FIRST FLOOR PLAN

TEST DRAWING 9B
FIRST FLOOR PLAN ELECTRICAL

TEST DRAWING NO. 9C (See pages 153-156)

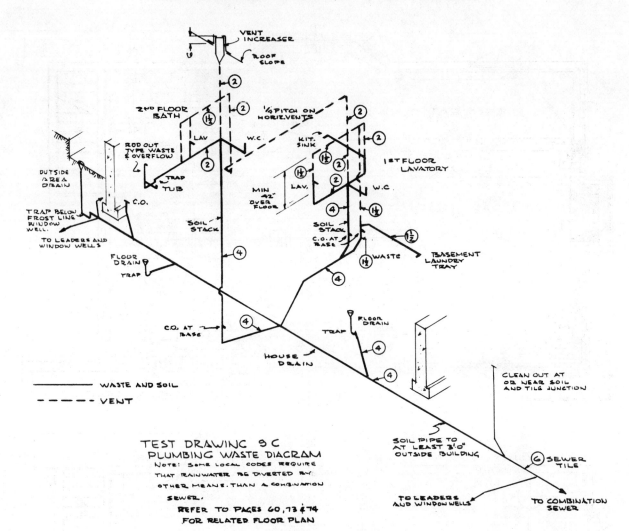

Reading Blueprints for Trade Information 151

TEST DRAWINGS NO. 9D AND 9E (See pages 153-156)

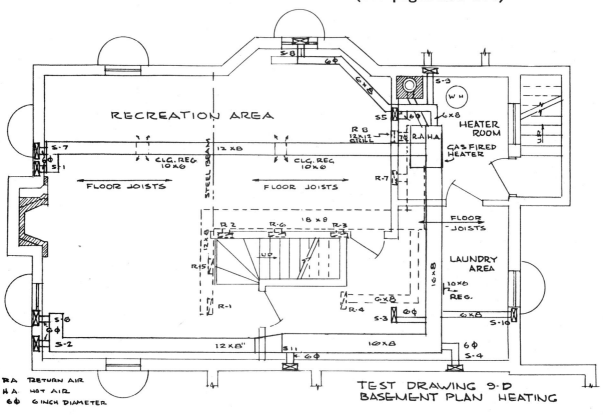

TEST DRAWING 9-D
BASEMENT PLAN   HEATING

RA RETURN AIR
HA HOT AIR
6∅ 6 INCH DIAMETER

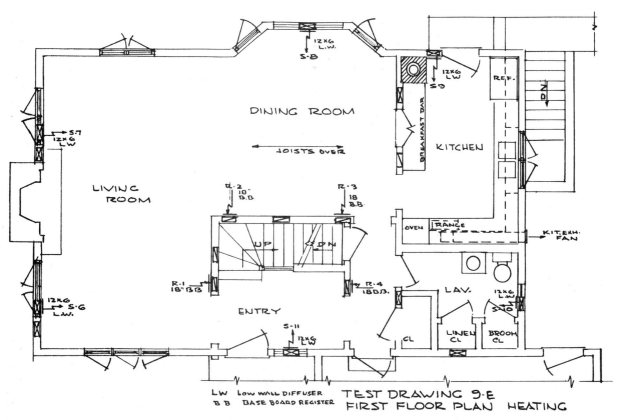

LW Low Wall Diffuser
BB Base Board Register

TEST DRAWING 9-E
FIRST FLOOR PLAN   HEATING

152 Building Trades Blueprint Reading

TEST DRAWINGS NO. 9F AND 9G (See pages 153-156)

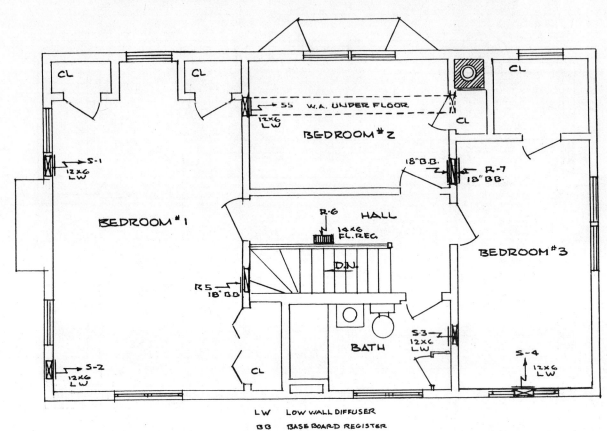

TEST DRAWING 9F
SECOND FLOOR PLAN HEATING

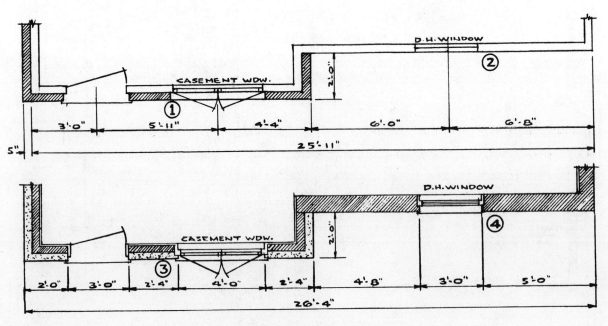

TEST DRAWING 9G

Trade Competency Test No. 9

(Chapter 10: Based on Test Drawings No. 9A, 9B, 9C, 9D, 9E, 9F, 9G, and Figures in the text.)

STUDENT'S NAME	INSTRUCTOR'S NAME

MULTIPLE CHOICE TEST: The following statements are incomplete or are in the form of a question to be answered. Choose the item which will best complete the statement or answer the question and place the letter in the space provided.

ELECTRICAL WORK

Refer to Test Drawings 9A and 9B

1. Referring to Test Drawing 9A Basement Plan—Electrical
    A. The power drop at the meter indicates that the service is 120 volts.
    B. There are six separate circuits in the basement.
    C. All circuits shown on this drawing are complete.
    D. Ceiling light outlets and convenience outlets are placed in separate circuits.

2. Referring to Test Drawing 9B First Floor Plan Electrical
    A. There are six complete circuits on the first floor.
    B. No circuit has more than nine lighting or convenience outlets.
    C. Three conductors are required in the conduit at each switch.
    D. Exterior lighting outlets are on individual circuits.

3. In Circuits 1 to 4
    A. In Circuit No. 1 (Test Drawing 9A) the ceiling lights in the laundry area are controlled by the switch at the inside stairs.
    B. Circuit No. 2 is a 240 volt outlet serving the washer and dryer.
    C. Circuit No. 3 is a separate circuit for convenience outlets which might be used for special appliances such as an iron or freezer.
    D. Circuit No. 4 serves light outlets and convenience outlets which are controlled by a switch.

4. In Circuits 5 to 7
    A. Circuit No. 5 contains only one outlet which supplies current to the furnace motor and controls.
    B. Circuit No. 6 contains three lighting outlets, one convenience outlet and two switches.
    C. In Circuit No. 7 (Test Drawing 9B) all of the convenience outlets are split wired and are controlled by a switch at the stairs.
    D. The 3-way switch located at the inside stairs with a lead to the second floor is part of circuit No. 7.

(Cont.)

Trade Competency Test No. 9 (Cont.)

5. In circuits 8 and 9
    A. Three conductors are required in all of the conduit runs in Circuit No. 8.
    B. Convenience outlets in the garage are controlled by 3-way switches.
    C. Circuit No. 9 includes nine lighting and convenience outlets.
    D. The location of all the switches in Circuit No. 9 are shown on the First Floor Plan—Electrical (Test Drawing 9B).

6. In Circuits 10 and 11
    A. Circuit No. 10 only includes dining room outlets, convenience outlets for lights on the oven and range hood and other light outlets in the kitchen.
    B. The ceiling light near the center of the kitchen is controlled by switches at three locations.
    C. The 3-way waterproof switch at the outside stairway is included in circuit No. 11.
    D. Circuit No. 11 is intended primarily for convenience outlets which might be used for kitchen appliances including the refrigerator, mixers, toasters etc.

PLUMBING

Refer to Test Drawing 9C

7. Regarding vertical soil waste and vent stacks
    A. The main soil stack continues as a vent extending through the roof with a 4 inch pipe all the way.
    B. The soil stack for the water closet in the first floor lavatory continues as a vent up through the roof.
    C. The waste and vent stack for the laundry trays is a 1½ inch pipe.
    D. A separate waste stack is provided to the house drain for the first floor lavatory.

8. Regarding horizontal vent and soil piping
    A. Horizontal vent piping is made perfectly level.
    B. The horizontal waste and soil piping at the first floor lavatory and the bathroom are 2 inch pipe.
    C. Horizontal vent pipes in the first floor lavatory and the bathroom partitions are placed approximately three feet above the floor.
    D. The kitchen sink waste passes horizontally to the waste stack which serves the laundry trays.

(Cont.)

Trade Competency Test No. 9 *(Cont.)*

Tear Off Here

9. Regarding waste and vent connections at fixtures
    A. The kitchen sink is vented through the vent system for the first floor lavatory.
    B. Waste from the tub enters the horizontal waste pipe without being trapped.
    C. The lavatory in the second floor bathroom has separate waste and vent pipes.
    D. Both water closets have 4 inch soil stacks and 4 inch waste stacks.

   HEATING

   Refer to Test Drawings 9D 9E and 9F

10. Referring to Test Drawing 9D
    A. All of the supply risers and return air drops to and from all floors are indicated.
    B. Return air ducts are all located in the vicinity of the interior stairway.
    C. All of the ducts are run between joists rather than below them.
    D. All of the ducts are rectangular in shape.

11. Referring to Test Drawings 9E
    A. A cold air return is shown in the kitchen.
    B. There are five cold air returns in the living room-dining room area.
    C. The duct shown in the wall of the linen closet goes to the second floor.
    D. The ducts shown in the breakfast bar are both warm air.

12. Referring to Test Drawing 9F
    A. All warm air diffusers are placed on outside walls.
    B. Bedrooms 2 and 3 share the same return duct.
    C. The supply duct for bedroom 2 rises vertically from the basement.
    D. Cold air registers are all baseboard units.

13. Regarding supply ducts
    A. Duct for S-2 is the longest run.
    B. The laundry has no direct heat supply.
    C. Supply outlet S-3 heats the entry.
    D. The recreation room is heated by supply registers S-6 and S-7.

	9.		
	10.		
	11.		
	12.		
	13.		

(Cont.)

Trade Competency Test No. 9 (Cont.)

14. Regarding return ducts
    A. There is no cold air return from the recreation room.
    B. There are seven cold air returns in all.
    C. The return air duct for the laundry is designated S-4.
    D. Warm air from S-6, S-7, and S-8 is exhausted through R-1, R-2, and R-3.

CONCRETE WORK AND MASONRY

Refer to Figures on pages 83 & 84, Figures 10-1, 10-37, 10-39 and Test Drawing 9G.

15. Regarding formwork for the foundation of the House Fig. 10-39
    A. All formwork is erected using 2 x 8 and 4 x 8 foot panels with fillers of various widths.
    B. Two foot wide panels or fillers are never used at corners.
    C. Double wales are only used on the outside of the forms.
    D. Provisions are made for window and door frames when the formwork is erected.

16. In Test Drawing 9G
    A. All walls have plaster or drywall inside finish.
    B. 1 is stone veneer on cement block.
    C. 2 is ordinary construction.
    D. 3 is stone veneer or brick.

17. Regarding dimensioning and locating window and door frames
    A. Special care must be taken in measuring for the window location in the brick part of a brick veneer house.
    B. Dimensions for openings for a house of solid brick construction is given to the sides of brick openings.
    C. Bricklayers are responsible for locating and setting window and door frames.
    D. All openings shall be laid out on a modular scale.

18. Regarding masonry veneer
    A. Brick veneer over a wood frame structure requires an air space of approximately one inch.
    B. The brick veneer part is not attached to the wood frame part.
    C. Ventilation of a brick veneer is optional.
    D. Stone veneer over brick is called ordinary construction.

19. The modular idea
    A. Permits the use of any type of brick.
    B. Requires a module based on a 6 inch grid.
    C. Is based on a 4 inch grid.
    D. Involves only horizontal measurements.

SCORE:

Chapter **11**

Reading a Set of Blueprints

The task of reading blueprints is a means toward an end. The craftsman is interested in reading the prints so that he may follow them in building the structure exactly as intended by the architect. A keen understanding of every detail of the blueprints helps him plan his work so that he may use his time and the material to the greatest advantage. The building of any house is a challenge, but it is a particular delight when the house is well planned. The craftsman shares in the pride of creating something beautiful and useful.

The blueprints provided at the end of the book qualify as worthy of careful analysis. The advertising literature supplied by the Ponderosa Pine Woodwork Association begins with this quotation:

> Here is a home that offers the utmost in casual, carefree modern living . . . and made possible by a revolutionary new concept . . . Total Privacy Zoning. The interior is zoned so that formal areas are away from work areas . . . children away from adults . . . older children apart from the younger ones . . . noisy rooms from quiet areas. Everything is easy to get to yet set apart. And the artful, generous use of elegant Ponderosa Pine woodwork gives this home quiet dignity and warmth. All in all . . . inside and out . . . here is a home carefully designed with you in mind . . . with more privacy than you'd expect in a home costing twice as much.

The excellent planning done by the architect will become evident as the many features of the building are examined.

There are several ways to study a set of blueprints. In general however the best procedure is to study the blueprints with the purpose of (a) understanding the floor plans and (b) gaining an idea of what the house will look like. After a preliminary viewing of the plans as a whole, each sheet should be studied in detail.

The logical starting point is the front entry on the *Upper Level Floor Plan, Sheet 2.* (Fig. 11-1 shows Sheet 2 in sketch form.) The living room, dining room, and master bedroom are arranged to be accessible yet private. The kitchen is placed so as to be the center of activity and control. It is open to the patio on one side and to the family room on the other side with a view through the family room windows to the playcourt beyond. Stairs ascend to Bedrooms 1 and 2 and Bathroom 2, which are one-half story above the other rooms shown on this sheet.

Lower Level Floor Plan, Sheet 1, is the other floor plan. (Fig. 11-2 shows Sheet 1 in sketch form.) Stairs descend from the family room to the level of Bedrooms 3 and 4, and Bathroom 3. These rooms are directly below Bedrooms 1 and 2, and Bathroom 2. Stairs descend again to the basement level. The basement extends to the north end of the

Note: Through the courtesy of the Ponderosa Pine Woodwork Association, an organization made up of many manufacturers of wood building products, the American Technical Society was granted permission to use the set of blueprints which appears at the end of the book. The building was designed by one of the leading architects of the country, G. Hugh Tsuruoka. It was used in a national campaign in conjunction with local lumber dealers to focus attention on the use of Ponderosa pine woodworking in fine houses. The American Technical Society is grateful for the opportunity to use the blueprints of this house.

158 Building Trades Blueprint Reading

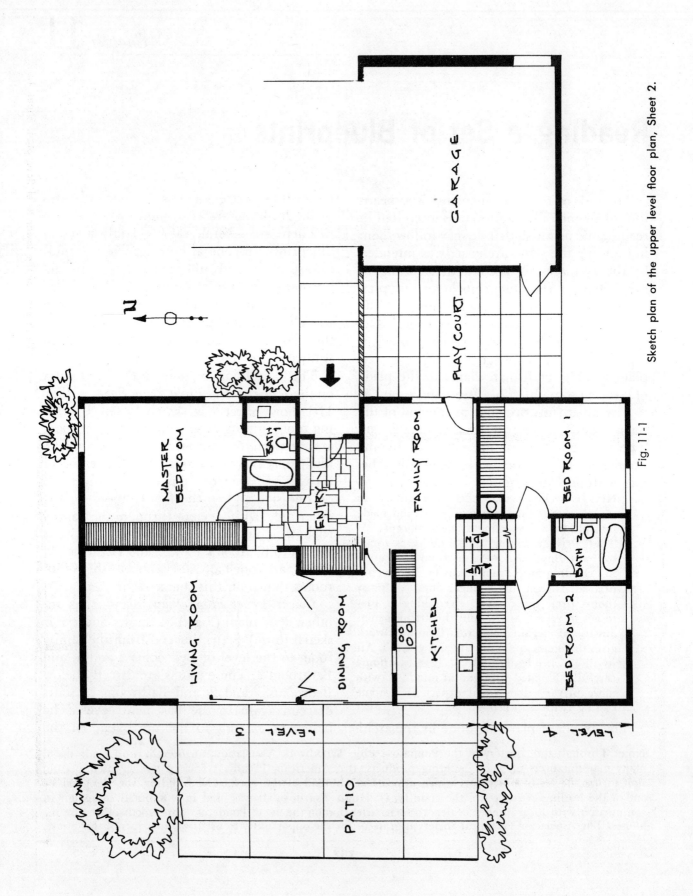

Fig. 11-1 Sketch plan of the upper level floor plan, Sheet 2.

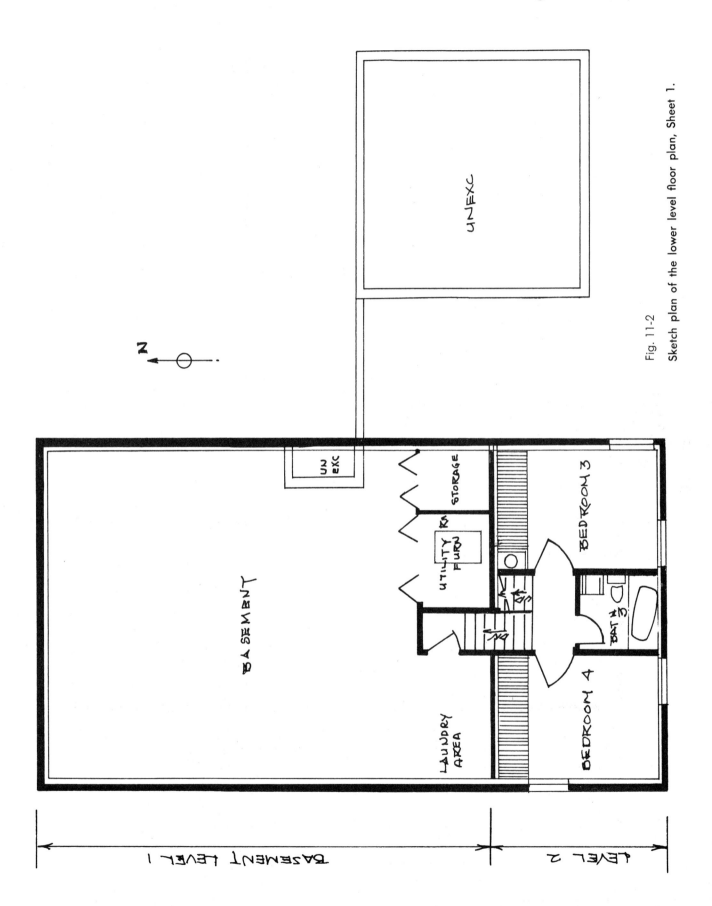

Fig. 11-2 Sketch plan of the lower level floor plan, Sheet 1.

160 Building Trades Blueprint Reading

house. Utility, storage, and laundry areas are located on this lower level.

The *East and West Elevations, Sheet 3,* and the *North and South Elevations, Sheet 4,* give views of the exterior. The general character of the house is contemporary. Casement windows are used with interesting vertical panel treatment. Large sliding glass doors with wood frames serve for light, ventilation, and access from the patio to the living room, dining room, and kitchen. This open arrangement is conducive to modern indoor-outdoor living. The roof slope is relatively low. Even though one part of the house extends almost two stories above the grade, the whole house appears to be built low to the ground. Fig. 11-3 and 11-4 show perspective views taken from two different vantage points.

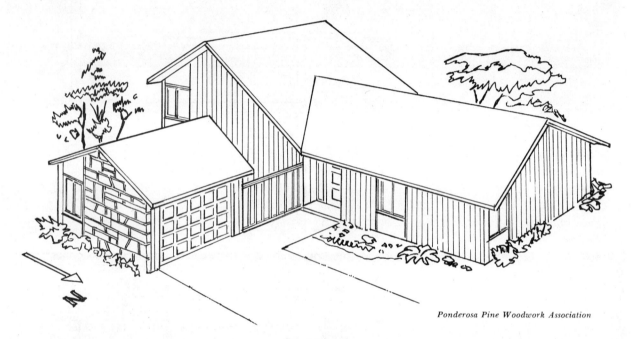

Ponderosa Pine Woodwork Association

Fig. 11-3

The house viewed from the north-east.

Ponderosa Pine Woodwork Association

Fig. 11-4

The house viewed from the south-east.

Reading a Set of Blueprints **161**

The *Longitudinal Section, Sheet 5,* is a view of an imaginary slice taken from end to end near the center of the house. It shows everything in section form which the cutting plane slices, and also shows what is seen beyond the part which is cut away. Thus details of construction of the foundation, columns, joists, beam, and rafters are revealed. Also the walls, doors, doorways, and openings, which would appear if the near part were removed, are indicated in their relative positions.

The longitudinal section shows how the four levels of the house are arranged. Beginning at the basement, each floor level is 4 feet 6 inches higher than the one below it.

Fig. 11-5 shows a cutaway view in perspective taken from the family room, with walls removed in order to see the stairs. The kitchen is in the far distance and the door to the right leads to the dining room. Stairs ascend to the upper level. The doors in the figure at this level are for Bedroom 2 and Bathroom 2. Another flight of stairs descends to a lower level. The doors shown at this lower level are for Bedroom 4 and Bathroom 3. This cutaway figure shows how the bedroom areas are set aside in a "Total Privacy Zone" yet are easily accessible to the family room and kitchen.

A Detailed Study; Upper Level Floor Plan, Sheet 2: *Entry.* The entry gives direct access to the living room, family room, and master bedroom. Daylight is provided by fixed glass sidelights at the front entrance door. The entry has an all-weather floor of slate. A generous coat closet is provided which has louvered bi-fold doors. A pocket door may be used to close off the family room when required. Exterior and interior lighting are controlled from a convenient location near the front entrance door.

Master Bedroom and Bath. The master bedroom is provided with casement windows on two walls for ventilation and view. The windows are not drawn with the usual conventional symbol because all of the windows in the house are alike and are shown on Sheet 6. A triple closet with louvered bi-fold doors extends along the west wall. The closets in this and the other bedrooms provide ample storage space. Because of their placement they serve to suppress noise from the adjacent rooms.

The bathroom has a counter type lavatory. Ventilation in all of the bathrooms is provided by fans.

The Living Room and Dining Room. These two rooms are arranged with a folding door-wall so that they can become one large room or be divided for privacy. The sliding glass doors, used in the place of windows, extend from the floor to a height of 6'-8". They open on to the patio and provide an atmosphere of indoor-outdoor living. See Fig. 11-6. A parquet floor adds interest. A parquet floor is made of strips of wood flooring glued to form tiles. They are laid in an alternating checkerboard fashion.

Kitchen. The kitchen is compact with cabinets and equipment arranged on each side of the room in corridor fashion. An open feeling of the outdoors is created by the large glass doors opening on the patio. Artificial light is

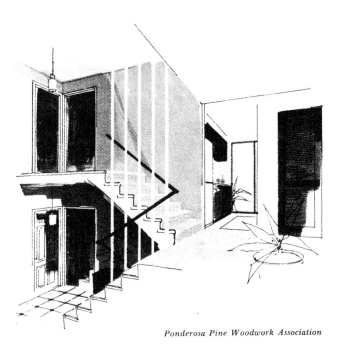

Ponderosa Pine Woodwork Association

Fig. 11-5

A cutaway view of the stairs shows how privacy is achieved in the bedroom areas.

Ponderosa Pine Woodwork Association

Fig. 11-6

The living room may be divided from the dining room by a folding door-wall. Glass doors open to the patio.

provided by a suspended luminous ceiling. The kitchen is provided with a pass-through opening to the dining room. This room is in close proximity to the playroom and the bedrooms and thus serves as the control center of the house. See Fig. 11-7.

Family Room. The family room is located so as to be accessible to the play court, to the kitchen, and to the bedrooms. It has a durable floor of composition tile.

Bedrooms 1 and 2 and Bathroom 2. Bedrooms 1 and 2 are adequate for children. Ample closets, ventilation, and wall space are provided. The bathroom is located between the two rooms. Note that the wall which contains the plumbing is 6 inches thick.

Lower Level Floor Plan, Sheet 1: *Bedrooms 3 and 4 and Bathroom 3.* Bedrooms 3 and 4 and Bathroom 3 are almost identical to those on the floor above. A study of the wall sections on Sheet 6 will show why these rooms are a few inches smaller because of the foundation wall. The chimney from the furnace is enclosed in a space in the north-west corner of Bedroom 3.

Basement. Stairs descend from the level of Bedrooms 3 and 4 to the basement level. The laundry is not shown in detail but is located below the kitchen to minimize plumbing piping. The utility room, containing the furnace and hot water heater, and the storage area are isolated from the basement by louvered bi-fold doors. Windows and areaways are not indicated on the blueprint, Sheet 1, because this house is for general application. They would be provided as desired by the owner and as required by the local building code.

Elevations, Sheets 3 and 4. At this point a study of the elevations in relation to the floor plans should be made to see which window on each elevation is a living room window, which is a master bedroom window, etc. Also a study of the floor levels should be made in relation to the Longitudinal Section View, Sheet 5.

The Elevation Sheets 3 and 4 are intended to show information about siding, window and door placement, elevations (levels) in relation to the grade, as well as information about cornices and other members of exterior trim. It takes much skill to place windows to serve the dual purpose of (a) respective room requirements and (b) enhancement of the structural appearance. The vertical treatment of the windows and the low pitched roof with a wide fascia add to the interest of the elevations. A stone veneer wall on the east side of the garage adds a strong accent. A wood fence with members placed in the position of vertical louvers sets off the play yard from the house and provides privacy.

Longitudinal Section, Sheet 5; and Wall Section, Sheet 6. The building is constructed using platform framing. The joists under the master bedroom and living room portion run east and west, resting on (a) a steel I-beam supported by steel pipe columns and (b) the foundation wall. The joists are hung so that the I-beam and the bottom of the joists are flush (or very nearly so). The joists under Bathroom 2 run in a north and south direc-

Ponderosa Pine Woodwork Association

Fig. 11-7

A feeling of being outdoors is conveyed by the artist in this picture of the kitchen. Note that the blueprints have not been followed precisely.

tion resting on a wall plate and a partition plate.

Wall Section at Bedrooms, Sheet 6. This shows information in detail about the footing, foundation, waterproofing, sheathing, siding, cornice, exterior and interior wall finish, etc. Notice that the outside of the concrete wall lines up with the face of studs. This is the clue that the outside dimensions of the building are to the face of studs.

Door and Window Types, Sheet 6. The door schedule should be studied in relation to the plan views in order to gain a clear understanding of where the doors are used. A dutch door, one which is divided so that the top can be opened independently of the bottom, is used to advantage in the family room.

Bathroom and Kitchen Elevations, Sheet 5. These interior elevation views are essential because they show much important information which would be difficult to indicate on a plan view. The arrangement of doors and drawers on kitchen cabinets, the location of the pass-through to the dining room, and information about the suspended luminous ceiling are examples. The bathroom elevations show the type of wall treatment, information about the dropped ceiling over the tub, location of mirrors, medicine chests, lighting, and details about the cabinet lavatory and the tub enclosure.

Final Examination

(Based on Blueprint Sheets 1 to 6)

STUDENT'S NAME

INSTRUCTOR'S NAME

MULTIPLE CHOICE TEST: The following statements are incomplete. Following each statement you will find four phrases which may be used to complete a sentence. Choose the phrase which will best complete the statement and place the letter in the answer column to the right.

	Answer	Score
1.		
2.		
3.		
4.		
5.		
6.		

1. The house including the basement is built on
    A. one floor level.
    B. two floor levels.
    C. three floor levels.
    D. four floor levels.

2. The roof of the house is
    A. an intersecting gable roof.
    B. a modified hip roof.
    C. a gable and shed roof.
    D. an ell shaped roof.

3. The roof on the house
    A. has eaves at the same level on the south and east sides.
    B. has all slopes equal (4.3 inches per foot of run).
    C. has ridges at the same level.
    D. has a different slope than the garage roof.

4. Materials used to cover the exterior walls of the house not including the garage
    A. consist of wood siding and stone veneer.
    B. consist of wood siding and exterior plywood.
    C. consist of wood siding, exterior plywood, and stone veneer.
    D. consist of boards and battens.

5. Windows
    A. are of different sizes.
    B. are casement with hopper sash below.
    C. are horizontal sliding.
    D. are outswinging and are all alike.

6. Exterior doors include
    A. a front entrance door with one light.
    B. a dutch door with louvers.
    C. a six-panel side door to the garage.
    D. a flush overhead garage door.

(Cont.)

Final Examination *(Cont.)*

7. The garage
    A. is attached to the house by a covered play court.
    B. is separated from the house by a play court with a fence at one end.
    C. has a level floor.
    D. has lighting, all of which may be controlled from the house.

8. Regarding floor levels,
    A. the basement floor level is 7'-6" below grade.
    B. the floor of bathroom 2 is 9'-0" above the basement floor.
    C. the patio slab is level with the first floor.
    D. the level of bathroom 2 is 4'-6" above grade.

9. The *Longitudinal Section, Sheet 5*
    A. is taken on a line passing through bathroom 2 and the living room.
    B. shows the kitchen opening, entry closet doorway, and living room opening.
    C. gives details of the construction of cornice at roof.
    D. shows partitions for bathrooms 2 and 3 directly above each other.

10. Regarding the structure,
    A. the joists under the living room run east and west.
    B. the steel I-beam passes under the joists.
    C. the steel I-beam extends the length of the building.
    D. joists under bathroom 2 run east and west.

11. Choose one of the following:
    A. the stairs marked "down" from the family room and "up" from bedroom 3 level are different stairs.
    B. bathroom 1 is above the unexcavated rectangle on the east wall in the basement.
    C. the laundry area is below the dining room.
    D. stairs marked "up" from kitchen level are directly over stairs to the basement.

12. The *Wall Section at bedrooms, Sheet 6* shows
    A. that the house is built with balloon framing with firestops at floor levels.
    B. that the bedrooms have lath and plaster walls.
    C. that the space under the roof is not vented.
    D. details as to how the foundation wall is to be covered on the room side.

(Cont.)

Final Examination *(Cont.)*

13. Insulation is provided
    - A. under the basement floor slab.
    - B. in exterior walls and over upper floor ceilings.
    - C. and to be of loose fill type.
    - D. under roof sheathing.

14. Regarding dimensions,
    - A. bedroom 4 is 10'-8" × 11'-4".
    - B. the basement is 29'-0" × 39'-0".
    - C. the exterior dimensions of the house are 29'-0" × 54'-0".
    - D. the play court is 12'-0" × 19'-4".

15. Interior doors
    - A. include six panel doors, louver doors, and louver doors with panels.
    - B. are all swinging or bi-fold.
    - C. to closets are all bi-fold.
    - D. All bi-fold doors are the same size.

16. Flooring in the house includes
    - A. wood parquet, composition, and ceramic tile.
    - B. slate, wood parquet, wood, and composition.
    - C. concrete, composition tile over concrete, and wood parquet over concrete.
    - D. composition tile in bedrooms 1, 2, 3 and 4.

17. Bathroom. Choose the best statement.
    - A. The master bath elevation and plan shows a counter top lavatory with electric lighting outlets at each side.
    - B. The ceiling in the tub enclosure in the master bath is 8'-0" above the floor.
    - C. A 1'-0" ceramic tile soffit is shown in bathroom 2.
    - D. In bathroom 2 ceramic tile is used only in the tub enclosure.

18. The *Kitchen Elevations, Sheet 5* show
    - A. that the ceiling height in the kitchen is 8'-0".
    - B. that doors are indicated on the pass-through to the dining room.
    - C. that the south wall shows cabinets, dishwasher, sink, and refrigerator-freezer.
    - D. that the north wall shows cabinets, range, range hood, pass-through, and a door to the dining room.

(Cont.)

19. Plumbing details include
    A. three floor drains in the basement.
    B. four hose bibbs.
    C. waste stack locations.
    D. plumbing partitions 6 inches thick.

20. Regarding electrical outlets,
    A. a waterproof outlet is provided for outdoor electrical needs.
    B. a 3-way switch in the family room controls all of the outside lights on the garage.
    C. outlets in the master bedroom are controlled by a switch.
    D. the luminous ceiling light in the kitchen is controlled by two switches.

21. Bathroom and plumbing fixtures include
    A. a laundry tray.
    B. three lavatories and water closets.
    C. three tubs which are all alike.
    D. medicine chests over each lavatory.

22. Regarding internal dimensions,
    A. the 4 inch partition thickness represents 3½ inches for the stud with ½ inch leeway.
    B. all partitions on Sheet 2 are shown as 4 inches thick.
    C. some dimensions on Sheet 1 are shown to the centerline of partitions or exterior walls.
    D. dimensions are given locating some interior doors from room corners on Sheet 2.

23. The stone veneer wall
    A. appears in the N, S, and E elevations.
    B. is built with a wood frame wall.
    C. is built with a concrete block wall.
    D. continues directly below garage window.

24. Regarding electrical outlets:
    A. ceiling outlets are provided in all bedrooms.
    B. play court lights are controlled by switches in the garage and family room.
    C. the two lights in the entry have separate switches.
    D. all wardrobes have light outlets

25. Regarding stairs:
    A. The difference in height from walk to entry is two risers.
    B. each tread on stairs from kitchen to bedroom 2 is 12" wide.
    C. stairs from kitchen to bedroom 2 has 8 treads.
    D. stairs from kitchen to bedroom 2 is closed off by doors.

SCORE:

Appendix A

REVIEWING MATHEMATICS

The ability to read blueprints intelligently involves a certain amount of mathematical skill. It is assumed that the student has the ability to work with whole numbers, fractions, and decimals in the mathematical processes of addition, subtraction, multiplication, and division. The Self-check Quiz on Mathematics at the end of this appendix can help to determine if he has need of special assistance in these areas. If so, practical mathematical textbooks are available for study.

Working With Inches and Feet. If you measured the piece of lumber shown in Fig. 1 and marked off two feet each time with your steel square, as indicated by the spaces

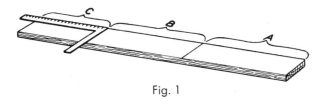

Fig. 1

marked *A* and *B,* and then found that the remaining length of lumber *(C)* was 1 foot 7 inches, what would be the total length of the piece of lumber?

```
2'  0"
2'  0"
1'  7"
─── ───
5'  7"  (Add the inch column first,
         then the foot column)
```

Ordinarily we prefer to reduce numbers to their simplest form. That is, instead of saying, for example, a board is 41 inches long, we would reduce this to feet. Three feet is 36 inches and subtracting 36 inches from 41 inches we have 5 inches left (41 − 36 = 5). We would say the board is three feet and five inches long, and we would write it 3'-5". The dimension of 41 inches can also be reduced to feet by dividing 41 by 12. We would then have 3 feet with 5 one inch units left over to give us 3'-5".

Here is a case where you add three dimensions together and then reduce inches to feet:

```
2'   6"
1'   7"
     4"
─── ───
3'  17"
4'   5"
```

By adding the *inch* column we get a total of 17 inches, which is more than one foot. Therefore it should be reduced to feet. 17 inches equals one foot and five inches (17 − 12 = 5). We would leave 5 in the *inch* column and add *1* to the *foot* column, giving us a total length of 4 feet and 5 inches.

Working With Fractions. Many people who use a ruler or framing square do not realize that they are constantly applying the basic rules of working with fractions. All of the rules can be worked out by observing the space between the inch marks of a ruler.

As you know, we use symbols and signs in mathematics as a shorthand to avoid writing many words. For example, when we divide 3 by 4 we write it ¾. The line between the 3 and 4 means divide. The number above the line is called the *numerator* and the one below the line the *denominator*.

Fig. 2 is an enlarged section of a ruler or

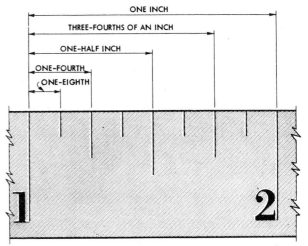

Fig. 2

framing square which shows the principal subdivisions. Notice that there are eight spaces in this inch that are the same length as the one marked *one-eighth*. In other words, the inch has been divided into 8 parts and each part would be $\frac{1}{8}''$ (one inch divided by 8). As there are 8 of these $\frac{1}{8}''$ spaces we can say:

$$\tfrac{1}{8}''+\tfrac{1}{8}''+\tfrac{1}{8}''+\tfrac{1}{8}''+\tfrac{1}{8}''+\tfrac{1}{8}''+\tfrac{1}{8}''+\tfrac{1}{8}''=1''$$
$$\text{or } \tfrac{8}{8}''=1''$$

Likewise there are two one-half inches

$$\tfrac{1}{2}''+\tfrac{1}{2}''=1'' \text{ or } \tfrac{2}{2}''=1''$$

and four one-fourth inches

$$\tfrac{1}{4}''+\tfrac{1}{4}''+\tfrac{1}{4}''+\tfrac{1}{4}''=1'' \text{ or } \tfrac{4}{4}''=1''$$

This gives us our first rule: *To add fractions with the same denominator (bottom number) add the numerators (top number).*

$$\text{So } \tfrac{1}{4}''+\tfrac{2}{4}''=\tfrac{3}{4}''$$

Now examine the space marked *one-half inch* in Fig. 2. It is made up of 2 *one-fourth inch* spaces or four *one-eighth inch* spaces, so we would write this

$$\tfrac{2}{4}=\tfrac{1}{2} \text{ and } \tfrac{4}{8}=\tfrac{1}{2}$$
$$\text{or } \tfrac{2}{4}=\tfrac{4}{8}=\tfrac{1}{2}$$

If we multiply both the numerator and denominator of $\frac{2}{4}$ by 2 we get $\frac{4}{8}$, and if we divide both the numerator and denominator of $\frac{4}{8}$ by *4* we get $\frac{1}{2}$. This brings us to our second rule for handling fractions:

The value of a fraction is not changed if we multiply or divide both the numerator and denominator of a fraction by the same number. For example,

Multiplying by 4 $\frac{1}{2}=\frac{4}{8}$ and $\frac{2}{3}=\frac{8}{12}$
Dividing by 2 $\frac{4}{8}=\frac{2}{4}=\frac{1}{2}$

Adding Fractions. Remember that in adding fractions we mentioned that they must have the same denominator. Now that we have Rule No. 2 we can add any fractions.

Suppose we wanted to add

$$\tfrac{1}{2}''+\tfrac{1}{4}''+\tfrac{1}{8}''$$

On the ruler we could figure it like this:

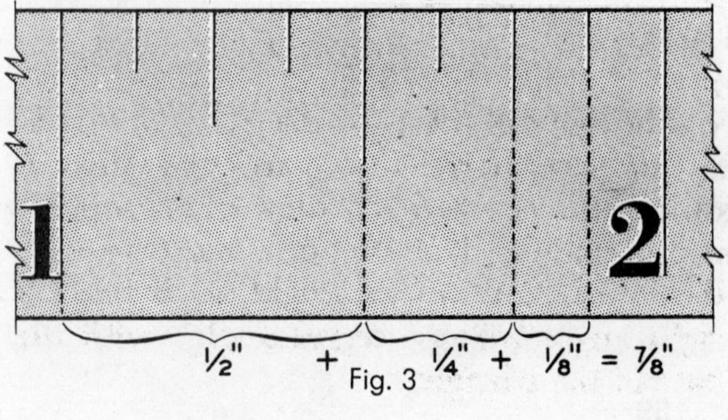

Fig. 3

Count the spaces and you will see the total is seven eighths, see Fig. 3, but rather than count we can easily do it by arithmetic.

Change each fraction to eighths, thus,

Add:	$\frac{1}{2}+\frac{1}{4}+\frac{1}{8}$
Change to *eighths.*	$\frac{1\times4}{2\times4}+\frac{1\times2}{4\times2}+\frac{1}{8}$
Add numerators:	$\frac{4}{8}+\frac{2}{8}+\frac{1}{8}=\frac{7}{8}$

In this case we changed all the denominators to *8*, which is the smallest number into which each denominator can be divided. In most cases it is easy to select what is known as the *Least Common Denominator* (L.C.D.) Check these examples carefully:

Add: $\frac{7}{8}+\frac{1}{16}=\frac{14}{16}+\frac{1}{16}=\frac{15}{16}$
Add: $\frac{1}{2}+\frac{1}{4}=\frac{2}{4}+\frac{1}{4}=\frac{3}{4}$
Add: $\frac{12}{16}+\frac{3}{32}=\frac{24}{32}+\frac{3}{32}=\frac{27}{32}$

The same rule applies to subtracting fractions: they must have the same denominator. Here are some examples :

Subtract: $\frac{7}{8}-\frac{1}{16}=\frac{14}{16}-\frac{1}{16}=\frac{13}{16}$
Subtract: $\frac{1}{2}-\frac{1}{4}=\frac{2}{4}-\frac{1}{4}=\frac{1}{4}$
Subtract: $\frac{3}{4}-\frac{3}{32}=\frac{24}{32}-\frac{3}{32}=\frac{21}{32}$

Working with Areas: Squares — Rectangles. A square is a four-sided figure in which all sides are equal in length and all the angles are *right* angles. Fig. 4 shows the floor plan of a kitchen that is 12 feet square, laid with tile that are 1 foot square.

Some materials such as plywood are sold by the *square foot* of area; other materials such as linoleum are sold by the square yard. A *square*

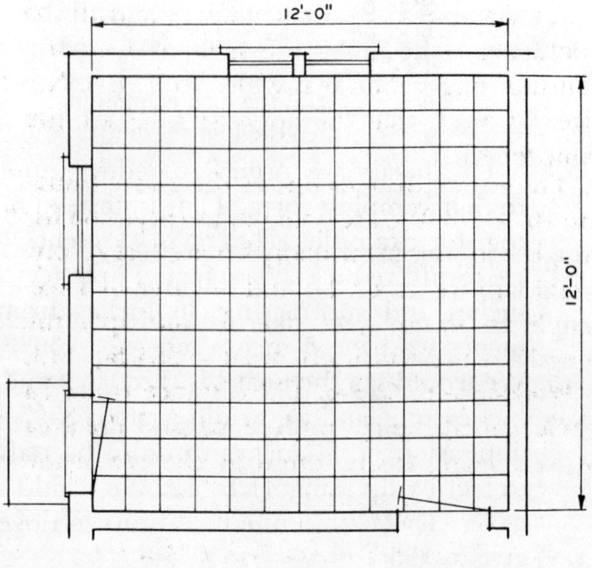

Fig. 4

foot is equal to the area of a square that is 1 foot on each side like the tile shown in this kitchen. If you count the squares, which are drawn to scale, you will find that there are 144; however, you can obtain this number easily by multiplying the length by the width (12 × 12 = 144 Sq. Ft.). Likewise if you wanted the number of square yards of linoleum needed for this kitchen you would change the width and length to yards (12 ft. = 4 yds.) and multiply (4 × 4 = 16 Sq. Yds.).

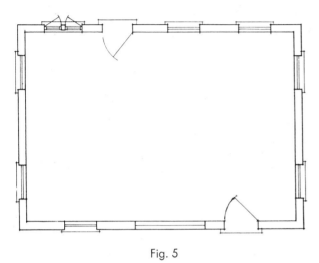

Fig. 5

Whenever we want the area of any surface we must use the same units. That is, to find square inches of area we must have our measurements in inches. If we want *square yards*, our measurements must be in yards, etc.

A rectangle is a four-sided figure and all the angles are right angles the same as a square, but it is longer than it is wide. See Fig. 5. Notice, however, that the opposite sides are the same length.

The inside dimensions of the above house are 16′ × 24′. We are going to use plywood for rough flooring. How many square feet of plywood will we need? To find the area of a rectangle in *square feet*, merely multiply the length by the width (16 × 24 = 384 Sq. Ft.).

Working with Circles. The distance around a circle, called the circumference, and the area of a circle are easily found by the use of the following two equations:

$$\text{Circumference} = D \times 3.1416$$
$$\text{Area} = 3.1416 \times r^2$$

The number 3.1416 is always used in finding the area or circumference of a circle. Because it is always constant, it is generally identified by the Greek letter π (pronounced pie). The r in the formula stands for the radius of the circle and is one-half the diameter (D) of the circle.

Examples: How long a steel band will be required to go around a column 4′ in diameter?

Length of steel band (circumference) =
4 × 3.1416
Length of steel band = 12.5664′
.5664 × 12 = 6.80 inches
The band is 12′-6 13/16″.

What is the area of a circle that has a diameter of 8′?

$$\text{Area} = 3.1416 \times r^2$$
$$" = 3.1416 \times 4^2$$
$$" = 3.1416 \times 16$$
$$" = 50.2656 \text{ sq. ft.}$$

or approximately 50¼ sq. ft.

Working with Triangles. Some roof trusses are good examples of triangles. Fig. 6 is a sketch of the nailed-glued trusses which have come into use quite generally in recent years because of their economy and strength.

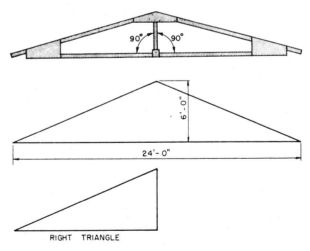

RIGHT TRIANGLE

Fig. 6

The shape of this truss is a *triangle;* that is, it has three sides. This particular truss is called a *king-post truss* because of the support in the middle. The *king-post* is square with the lower chord, or we might say it forms a right angle (ninety degree angle) with the base of the triangle.

When it is necessary to find the area of a triangle the formula to use is:

$$\text{Area} = \frac{\text{altitude} \times \text{base}}{2}$$

The area for a gable on the house which uses the truss in Fig. 6 would be calculated thus:

$$\frac{6' \times 24'}{2} = 72 \text{ sq. ft.}$$

Pitch of Roof. Pitch is expressed as a *fraction*—the ratio of the rise (*distance of plate to ridge*) to the width (*span*).

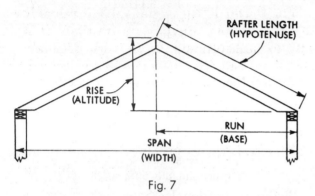

Fig. 7

Stated as a simple formula that is easy to remember, it would be:

$$P \text{ (pitch)} = \frac{R \text{ (rise)}}{S \text{ (span)}}$$

So if the rise of a roof is 4' and the span 24', then

$$P = \frac{4}{24} = \frac{1}{6}$$

Common roof pitches are 1/6, 1/4, 1/3, 1/2, 2/3, 3/4, and 1.

Unit Measurements. In order to simplify roof measurements we generally refer to the *unit rise,* which is the amount of rise for each foot of *run.*

Therefore, if we have a rise of 4' (48 inches) and the run is 12 feet (one-half the span), we would say that the unit rise for each foot of run is 4 inches ($48/12 = 4$) and this unit measurement is generally shown in the elevation view as in Fig. 8.

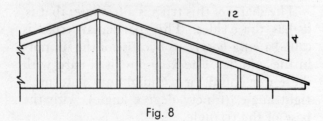

Fig. 8

If you have the unit rise and run, and want to find the pitch, use the formula for Pitch.

$$P = \frac{R}{S}$$

$$P = \frac{4}{12 \times 2} \text{ (span is twice the run)}$$

$$P = \frac{4}{24} = \frac{1}{6}$$

Following is a list of common roof pitches with their corresponding unit measurements:

Pitch	Unit Run	Unit Rise
1/6	12″	4″
1/4	12″	6″
1/3	12″	8″
1/2	12″	12″
2/3	12″	16″
3/4	12″	18″
1 (full pitch)	12″	24″

Length of Rafter. In order to find the length of a rafter arithmetically, we use the Pythagorian theorem which you will remember from Junior High School mathematics: *The square of the hypotenuse of a right triangle is equal to the sum of the squares of the other two sides.*

This can be stated as follows (refer to Fig. 7):

$$H^2 = B^2 + A^2$$

H is hypotenuse or rafter length
B is base or run
A is altitude or rise

In order to find the length of the rafter if the run is 14' and the rise is 7' we would substitute in the above equation:

$$H^2 = 14^2 + 7^2$$
$$H^2 = 196 + 49$$
$$H^2 = 245$$
$$H = \sqrt{245} \text{ or } 15.65'$$

$.65 \times 12 = 7.80$ inches

Rafter Length 15'-7 13/16″

In case you have forgotten how to extract square roots the following example should refresh your memory:

```
                              1 5. 6 5
                            ┌─────────────
                            │ 2'45.00'00
Largest square is .......  1│ 1
                            │ ───
Trial divisor (20 × 1) .. 20│ 1 45
Complete divisor ......   25│ 1 25
                            │ ─────
Trial divisor (20 × 15) . 300│   20 00
Complete divisor ......  306│   18 36
                            │   ──────
Trial divisor (20 × 156) 3120│    1 64 00
Complete divisor ...... 3125│    1 56 25
                            │    ───────
                            │        7 75
```

Board Foot. Strip lumber, boards, dimension lumber, structural timbers, and shop lumber are sold by the *board foot*. One board foot is 1 inch thick, 12 inches wide, and one foot long. We ordinarily designate a board by its thickness and width in inches and its length in feet. For example, we say a <u>two by four</u> <u>ten feet long</u>. Likewise in figuring board feet we always multiply the thickness and width in inches, by the length in feet, and divide by 12, thus

$$\frac{\text{thickness (in.)} \times \text{width (in.)} \times \text{length (ft.)}}{12}$$

Examples:

How many board feet in a 2 × 4 that is 12 feet long?

$$\frac{2 \times 4 \times \cancel{12}^{1}}{\cancel{12}} = 8 \text{ board feet}$$

How many board feet in twelve 2″ × 8″ × 24′?

$$\frac{2 \times 8 \times \cancel{24}^{2}}{\cancel{12}} = 32 \text{ board feet in 1 board,}$$

$12 \times 32 = 384$ board feet in 12 boards.

Using Decimal Equivalents. The Decimal Equivalent Table (Fig. 9) comes in handy when calculations of problems do not come out in even inches. The Decimal Equivalent Table helps us find the nearest fraction of an inch. In the rafter length problem the length was found to be 15′-7.80″. Referring to the table, the fraction which we could use would be either $^{51}/_{64} = .797$ or $^{13}/_{16} = .812$. Rafters are not cut to the 64th of an inch, thus $^{13}/_{16}$ was chosen. The dimension to use would be 15′-7$^{13}/_{16}$″.

The Decimal Equivalent Table is valuable in working with problems in fractions. It is easy to multiply and divide with decimals, and if we can convert fractions to decimals we simplify our work.

For example, 29¼ × 5⅜ can be expressed as

$$\begin{array}{r} 29.25 \\ \times\ 5.375 \end{array}$$

which equals 157.218. Using the table to convert decimals back to fractions, this would be 157$^{7}/_{32}$.

4ths	8ths	16ths	32nds	64ths	to 2 places	to 3 places	to 4 places	4ths	8ths	16ths	32nds	64ths	to 2 places	to 3 places	to 4 places
				1/64	0.02	0.016	0.0156					33/64	0.52	0.516	0.5156
			1/32		0.03	0.031	0.0312				17/32		0.53	0.531	0.5312
				3/64	0.05	0.047	0.0469					35/64	0.55	0.547	0.5469
		1/16			0.06	0.062	0.0625			9/16			0.56	0.562	0.5625
				5/64	0.08	0.078	0.0781					37/64	0.58	0.578	0.5781
			3/32		0.09	0.094	0.0938				19/32		0.59	0.594	0.5938
				7/64	0.11	0.109	0.1094					39/64	0.61	0.609	0.6094
	1/8				0.12	0.125	0.1250		5/8				0.62	0.625	0.6250
				9/64	0.14	0.141	0.1406					41/64	0.64	0.641	0.6406
			5/32		0.16	0.156	0.1562				21/32		0.66	0.656	0.6562
				11/64	0.17	0.172	0.1719					43/64	0.67	0.672	0.6719
		3/16			0.19	0.188	0.1875			11/16			0.69	0.688	0.6875
				13/64	0.20	0.203	0.2031					45/64	0.70	0.703	0.7031
			7/32		0.22	0.219	0.2188				23/32		0.72	0.719	0.7188
				15/64	0.23	0.234	0.2344					47/64	0.73	0.734	0.7344
1/4					0.25	0.250	0.2500	3/4					0.75	0.750	0.7500
				17/64	0.27	0.266	0.2656					49/64	0.77	0.766	0.7656
			9/32		0.28	0.281	0.2812				25/32		0.78	0.781	0.7812
				19/64	0.30	0.297	0.2969					51/64	0.80	0.797	0.7969
		5/16			0.31	0.312	0.3125			13/16			0.81	0.812	0.8125
				21/64	0.33	0.328	0.3281					53/64	0.83	0.828	0.8281
			11/32		0.34	0.344	0.3438				27/32		0.84	0.844	0.8438
				23/64	0.36	0.359	0.3594					55/64	0.86	0.859	0.8594
	3/8				0.38	0.375	0.3750		7/8				0.88	0.875	0.8750
				25/64	0.39	0.391	0.3906					57/64	0.89	0.891	0.8906
			13/32		0.41	0.406	0.4062				29/32		0.91	0.906	0.9062
				27/64	0.42	0.422	0.4219					59/64	0.92	0.922	0.9219
		7/16			0.44	0.438	0.4375			15/16			0.94	0.938	0.9375
				29/64	0.45	0.453	0.4531					61/64	0.95	0.953	0.9531
			15/32		0.47	0.469	0.4688				31/32		0.97	0.969	0.9688
				31/64	0.48	0.484	0.4844					63/64	0.98	0.984	0.9844
1/2					0.50	0.500	0.5000	1					1.00	1.000	1.0000

Fig. 9   *American National Standards Institute*

DECIMAL EQUIVALENTS

When decimals with two places are used the answers are approximately correct. Greater accuracy results when three or four decimal places are used.

Self-Check Quiz on Mathematics

(Based on pages 169 to 173)

Solve the following problems without referring to the answer sheet. If too many errors are made or the process is not understood, additional study is suggested. *Answers are given in the appendix.*

Answer

WHOLE NUMBERS
1. Add   666 + 434 + 204 + 989   1. _____
2. Add   79 + 78892 + 6291 + 3   2. _____
3. Subtract   681 from 993   3. _____
4. Subtract   847 from 9321   4. _____
5. Multiply   784 by 13   5. _____
6. Multiply   6492 by 931   6. _____
7. Divide   3172 by 152 (to nearest whole number)   7. _____
8. Divide   74589 by 382 (to nearest whole number)   8. _____

FRACTIONS
9. Add $3/8 + 5/8 + 7/8$ 9. _____
10. Add $5/8 + 11/16 + 1/4$ 10. _____
11. Add $10\ 9/16 + 4\ 1/8 + 2\ 1/2 + 3$ 11. _____
12. Subtract $1\ 7/8$ from $2\ 3/8$ 12. _____
13. Subtract $7\ 13/16$ from $10\ 1/4$ 13. _____
14. Multiply $7/8 \times 3/4$ 14. _____
15. Multiply $2\ 1/4 \times 1\ 3/16$ 15. _____
16. Multiply $1\ 1/2 \times 1\ 1/2 \times 1\ 1/2$ 16. _____
17. Divide   21 by $1/4$ 17. _____
18. Divide $1/4$ by 21   18. _____
19. Divide $3\ 1/16$ by $1/4$ 19. _____

DECIMALS
20. Add   4.50 + 11.65 + 3.65 + .65   20. _____
21. Add   244.05 + 43.1 + 7.982 + 22   21. _____
22. Subtract   783.21 from 897.37   22. _____
23. Subtract   6476.201 from 7652.91   23. _____
24. Multiply   1125 × 1.5   24. _____
25. Multiply   76.5 × .25   25. _____
26. Multiply   1.321 × 2.51   26. _____
27. Divide   326.16 by 72   27. _____
28. Divide   39.375 by 4.375   28. _____
29. Divide   736.4 by 8.15 (3 decimal places)   29. _____
30. Divide   .133 by .7   30. _____

Appendix B

METRIC CONVERSIONS

Rapid expansion of trade and industry on an international basis in the past two decades has increased the need for understanding of both the *metric* or CGS (Centimeter-Gram-Second) system used by nearly all countries of the world and the *English* or FPS (Foot-Pound-Second) system used by the United States and some other English-speaking countries.

If the coexistence of two systems seems inconvenient, as it is, remember that in respect to worldwide agreement we are the exception. In view of the increasing need for a universal system to measure lengths, areas, volumes, weights, temperatures, etc., it now seems likely that the CGS system will ultimately replace the FPS system despite immense costs and problems that will be involved in making the changeover.

Existing rules and factors will have to be changed if the metric system is adopted. New measuring tools and devices will be introduced to replace existing ones in the construction trades. Measurements given in the metric system can be converted to the English system or *vice versa*. Actually, metric measurement should result in more accuracy because it eliminates multiplying and converting of fractions. A complete conversion to the metric system would, of course, take place over a long period of time.

In some areas, especially outside the United States, blueprints with dual dimensioning may be encountered: with *both* metric *and* English dimensions given. In still other areas, of course, the measurements are given only in the metric system. The craftsman can easily convert English measurements to metric and metric to English by using the instructions below with the tables on the following page.

Table 1 lists factors for converting units from metric to English, while Table 2 lists factors for converting from English to metric units.

To convert a quantity from *metric* to *English* units:

1. Multiply by the factor shown in Table 1.
2. Use the resulting quantity "rounded off" to the number of decimal digits needed for practical application.
3. Wherever practical in semi-precision measurements, convert the decimal part of the number to the nearest common fraction.

To convert a quantity from *English* to *metric* units:

1. If the English measurement is expressed in fractional form, change this to an equivalent decimal form.
2. Multiply this quantity by the factor shown in Table 2.
3. Round off the result to the precision required.

Relatively small measurements, such as 17.3 cm, are generally expressed in equivalent millimeter form. In this example the measurement would be read as 173 mm.

(SEE PAGE 176 FOR METRIC TABLES)

METRIC-TO-ENGLISH AND ENGLISH-TO-METRIC CONVERSION TABLES

TABLE 1 CONVERSION OF METRIC TO ENGLISH UNITS

LENGTHS:		WEIGHTS:	
1 MILLIMETER (mm)	= 0.03937 IN. OR = 0.003281 FT	1 GRAM (g)	= 0.03527 OZ. (AVDP)
1 CENTIMETER (cm)	= 0.3937 IN.	1 KILOGRAM (kg)	= 2.205 LBS
1 METER (m)	= 3.281 FT OR 1.0937 YDS	1 METRIC TON	= 2205 LBS
1 KILOMETER (km)	= 0.6214 MILES	LIQUID MEASUREMENTS:	
AREAS:		1 CU CENTIMETER (cc)	= 0.06102 CU IN.
1 SQ MILLIMETER	= 0.00155 SQ IN.	1 LITER (= 1000 cc)	= 1.057 QUARTS OR 2.113 PINTS OR 61.02 CU INS.
1 SQ CENTIMETER	= 0.155 SQ IN.		
1 SQ METER	= 10.76 SQ FT OR 1.196 SQ YDS	POWER MEASUREMENTS:	
		1 KILOWATT (kw)	= 1.341 HORSEPOWER
VOLUMES:		TEMPERATURE MEASUREMENTS:	
1 CU CENTIMETER	= 0.06102 CU IN.	TO CONVERT DEGREES CENTIGRADE TO DEGREES FAHRENHEIT, USE THE FOLLOWING FORMULA: DEG F = (DEG C X 9/5) + 32	
1 CU METER	= 35.31 CU FT OR 1.308 CU YDS		

SOME IMPORTANT FEATURES OF THE CGS SYSTEM ARE:
1 CC OF PURE WATER = 1 GRAM. PURE WATER FREEZES AT 0 DEGREES C AND BOILS AT 100 DEGREES C.

TABLE 2 CONVERSION OF ENGLISH TO METRIC UNITS

LENGTHS:		WEIGHTS:	
1 INCH	= 2.540 CENTIMETERS OR 25.40 MILLIMETERS	1 OUNCE (AVDP)	= 28.35 GRAMS
1 FOOT	= 30.48 CENTIMETERS OR 304.8 MILLIMETERS	1 POUND	= 453.6 GRAMS OR 0.4536 KILOGRAM
1 YARD	= 91.44 CENTIMETERS OR 0.9144 METERS	1 (SHORT) TON	= 907.2 KILOGRAMS
		LIQUID MEASUREMENTS	
1 MILE	= 1.609 KILOMETERS	1 (FLUID) OUNCE	= 0.02957 LITER OR 28.35 GRAMS
AREAS:		1 PINT	= 473.2 CU CENTIMETERS
1 SQ IN.	= 6.452 SQ CENTIMETERS OR 645.2 SQ MILLIMETERS	1 QUART	= 0.9463 LITER
1 SQ FT	= 929.0 SQ CENTIMETERS OR 0.0929 SQ METER	1 (U.S.) GALLON	= 3785 CU CENTIMETERS OR 3.785 LITERS
1 SQ YD	= 0.8361 SQ METER	POWER MEASUREMENTS	
VOLUMES:		1 HORSEPOWER	= 0.7457 KILOWATT
1 CU IN.	= 16.39 CU CENTIMETERS	TEMPERATURE MEASUREMENTS	
1 CU FT	= 0.02832 CU METER	TO CONVERT DEGREES FAHRENHEIT TO DEGREES CENTIGRADE, USE THE FOLLOWING FORMULA: DEG C = 5/9 (DEG F −32)	
1 CU YD	= 0.7646 CU METER		

Glossary

Anchors. Irons of special shape used to fasten masonry and wood parts together.

Anchor Bolt. A metal bolt used to tie down a wood sill to a masonry or concrete foundation wall.

Angle (Angle Iron). A piece of structural iron formed with the cross section shape of a right angle.

Apron. A piece of inside window trim placed under the stool, flat against the wall.

Areaway. An opening adjacent to a basement window or door to permit air and light to enter.

Ash Drop (or Ash Dump). A trap door for ashes in the floor of the fireplace leading to a chute.

Baseboard. A finish board covering the wall where it meets the floor.

Base Shoe. A small molding applied to the baseboard at the floor.

Batt Insulation. Blanket insulation cut into short lengths to make it easy to handle and apply.

Batten. A narrow strip of wood used to cover the joint between two vertical pieces of siding.

Batterboard. A construction of stakes and horizontal boards from which chalklines are hung which define the building lines.

Bay. A part of a room which projects out beyond the plane of the wall.

Bay Window. A window projecting outward from the face of a wall.

Beam. An iron structural member or heavy wood member used to support floor joists.

Bench Mark. A mark on some object firmly fixed in the ground from which distances and elevations are measured.

Bond. The arrangement of bricks in a wall.

Bridging. Crossbracing between joists and studs to add stiffness to the floor and walls.

Built-up Roof. A roof made up of several layers of felt, each spread with hot coal tar pitch or asphaltum.

Casement Window. A window in which each sash opens outward on hinges placed at the side of the sash.

Casing. A wood trim member covering the space between the plaster or drywall and the jamb at windows and doors.

Channel. A rolled piece of structural iron with sides bent up to take the cross section shape of a channel.

Chord. The lower horizontal member of a truss.

Cleanout. *(Masonry)* An opening at the foot of a chimney or fireplace to remove ashes and debris. *(Plumbing)* An angle pipefitting which is placed where necessary in a house drain so that it can be rodded out.

Component. A part of a house assembled before delivery to the building site.

Compressor. One of the main parts of an air conditioning system required in the cooling cycle.

Concrete. A mixture of sand, cement and gravel in varying amounts according to use, mixed with water.

Conduction. The transfer of heat by contact.

Conduit. Metal tubing used to carry electrical conductors.

Convection. A transfer of heat by a moving substance such as air or water.

Convector. A unit which is used to transfer heat from hot water pipes into the room.

Convenience Outlet. An electrical outlet in the wall which can be used for many purposes.

Cornice. A horizontal molded projection which crowns or finishes the eaves of a building.

Course. A level layer of brick, stone or other masonry material.

Cove Mold. A concave molding used on inside corners.

Curing. The chemical process which takes place in concrete after it is poured and as it attains its load bearing strength.

Damper. A moveable metal plate in a fireplace throat to regulate draft. A moveable metal plate in a duct to control the flow of air.

Datum Point. A point of reference established by the city from which levels and distances are measured.

Diffuser. An air register transferring forced air from a duct to the room.

Dormer. A projection, usually with a window, built out from a sloping roof.

Doublehung Window. A window with upper and lower sash which slide up and down in grooves of the window frame.

Downspout. A vertical pipe to carry rainwater from the gutter to the ground or sewer. (Same as leader.)

Drip Cap. A molding placed above the top of a window or door casing to provide a means for water to run off.

Drywall. A system of interior wall finish using sheets of gypsum board.

Duct. A large round or rectangular metal pipe used for carrying air.

Eaves. The portion of the roof which overhangs the wall.

Elevation. Drawings of buildings or parts of buildings made as though the observer were looking horizontally directly at the building.

Fascia. A flat vertical board located at the outer face of a cornice.

Firecut. A beveled cut made on the end of a joist which is inserted into a masonry wall.

Firestop. Any blocking of air passages to prevent the spread of fire in a building. (A block of wood closing off a space between studs.)

Flare. An operation performed on the end of copper tubing before it is joined to another piece.

Flashing. Sheet metal used in roof and wall construction to keep water out.

Flue. A passageway in a chimney for smoke and gas.

Footing. The spread portion at the base of a foundation wall or column which distributes the weight over a larger area.

Foundation. The part of the building usually below ground on which the building rests.

Frame Construction. Building using wood structural members.

Frieze. The part of a cornice which is the lower vertical board at the wall.

Frost Line. The depth to which frost penetrates the earth.

Furring Strips. Narrow strips of wood fastened to a wall or ceiling to serve as a leveling device and to provide a means for fastening the finishing materials.

Gable. The triangular end of a house with a gable roof.

Gable Roof. A ridged roof that slopes up from two sides.

Gambrel Roof. A roof with two different slopes on its two sides.

Girder. A horizontal structural member used to support floor joists.

Grade. The level of the ground around a building.

Gravel Stop. A strip of metal formed with a vertical lip which is used at the edge of a built-up roof.

Grout. A thin cement mortar.

Gusset. A flat surface of plywood or metal used to reinforce a joint of a truss.

Gutter. A horizontal rain trough at the edge of a roof.

Gypsum. A calcium product used in plaster and as a core in sheets of drywall.

Head. The top of a window or door frame.

Header. A joist or joists placed at the ends of an opening in the floor used to support side members. The top rough framing members over a window or door opening.

Hearth. Masonry apron in front of a fireplace.

Hickey. A tool used to bend conduit.

Hip Roof. A roof sloping up from all sides or walls of a building.

Hose Bibb. A water faucet or valve connection for a garden hose.

Hydronic System. Forced hot water system.

I Beam. A structural iron beam with a cross section resembling the letter I.

Isometric. A type of pictorial drawing in which horizontal lines in the object are drawn at a 30° angle from the horizontal.

Jalousie. A window consisting of narrow pieces of glass arranged horizontally which open outward.

Jamb. The main members of a window or door frame, forming the sides and top.

Jig. A table or device used to hold structural members in place while they are being fastened together. Used in the manufacture of components and trusses.

Joists. The framing members which directly support the floor.

Lally Column. A metal pipe filled with concrete used to support beams or girders.

Lath. Metal mesh or wood strips which are fastened to structural members to provide a base for plaster.

Laundry Tray. A laundry sink.

Lavatory. (1) A basin for washing the hands and face. (2) A room equipped with running water usually containing a basin and water closet.

Leader. A vertical pipe used to carry water from a roof gutter to the ground or to the sewer. (Same as downspout.)

Level. (1) Horizontal. (2) A builder's instrument used to transfer points in laying out foundations, revolves only in a horizontal plane. (3) A carpenter's or mason's tool used to level building parts in the course of building. (4) To adjust into a horizontal position.

Light. A pane of glass.

Lintel. A horizontal member supporting the wall over an opening.

Louver. A slatted opening used for ventilating attics and other roof spaces.

Louver Door. A door with horizontal ventilating slats.

Masonry. A means of building, using brick, stone, tile, terra cotta or concrete units.

Millwork. Finished woodwork machined and assembled in a mill such as windows, doors and their frames.

Modular. A structural system designed to have the parts fit together on a grid of a standard module.

Module. (1) A unit of measurement established at 4 inches. (2) A complete part of a building assembled in a shop such as a bathroom or kitchen.

Mullion. The structural member between windows which come in pairs or in a series.

Muntin. The small members dividing the glass lights in a window sash.

Nominal Size. The descriptive size of lumber, not the actual measured size. Usually 1×6, 2×4, etc.

Nosing. The extension on a stair tread.

Outlet. A point in an electric wiring system where current is drawn to supply a lighting fixture or appliance.

Parapet. A low wall at the edge of a roof.

Parting Strip. A vertical strip attached to the jamb of a double hung window used to separate the upper and lower sash as they slide up and down.

Partition. An interior wall which separates a space into rooms.

Pass-through. An opening in a kitchen wall used to pass dishes to and from the dining room. (Also *pass-thru*.)

Pitch. (1) The slope of a floor toward a drain expressed in inches per foot. (2) A ratio between the rise of a roof and the span.

Plancier. A board which is the underside of an eave or cornice.

Plaster. A pasty composition of lime, sand and water which hardens on drying, used for coating wall and ceiling surfaces.

Plate. The top horizontal structural member of a frame wall.

Plenum. An air compartment maintained under pressure connected to one or more distributing or return ducts.

Plumb. (1) Vertical. (2) To adjust into a vertical position.

Radiant Heat. A system in which space is heated by the use of pipe coils or electric resistance wires placed in the floor, ceiling or walls.

Radiation. The transfer of heat through space by the wave motion of rays.

Rafter. A sloping roof member which supports the roof covering.

Rail. A horizontal member of a door or a window sash.

Random. A manner of laying stones so that they do not follow regular patterns or courses.

Receptacle. A receptacle is a contact device installed at the outlet for the connection of a single attachment plug.

Receptacle Outlet. An outlet where one or more receptacles are installed.

Register. A fixture through which air flows into or out of a room for heating or cooling.

Ribbon. A narrow strip of board cut to fit into the edge of studding to help support joists.

Rise. On a roof built using *simple rafters,* the "rise" is the vertical (straight up and down and plumb) distance measured from the highest point (the ridge) of the rafter, to the lowest point at the other end of the rafter. See Appendix, "Reviewing Mathematics," Fig. 7.

Riser. A vertical board at the edge of a stairway step.

Rubble. A wall made of rough stones irregular in size and shape, laid without a pattern.

Run. *(Plumbing)* A part of a pipe or fitting that continues in the same straight line as the direction of flow. *(Roof)* On a roof built using *simple rafters,* the "run" is the horizontal (straight across and level) distance measured from the highest point (the ridge) of the rafter, to the farthest point at the other end of the rafter. See Appendix, "Reviewing Mathematics," Fig. 7.

Saddle. A small gable roof placed behind a chimney on a sloped roof to shed water.

Sash. The frame in which the window lights (glass) are set.

Scuttle. A small opening in the ceiling to provide access to the attic space.

Section. A view taken of a building or a part of a building as seen in a vertical slice.

Setback. A specified minimum distance that a structure must be placed from a lot line.

Septic Tank. A tank in which sewage is kept in order that bacterial action may cause disintegration of organic matter.

Sheathing. Fiberboard, gypsumboard, plywood or rough boards that cover up the superstructure and rafters.

Shiplap. Lumber which has been worked or rabbeted along each edge to provide a close lapped joint by fitting two pieces together.

Siding. Outside wall finish in long narrow units.

Sill. (1) The bottom rough structural member which rests on the foundation. (2) The bottom exterior member of a window or door.

Sill Cock. An outside valve for the connection of a hose.

Site. The location of the building.

Slab. A flat area of concrete.

Soffit. A lower horizontal surface such as the underface of eaves, cornice or beam.

Soil Stack. A vertical pipe which runs from the soil pipe to the house drain to carry waste including that from water closets.

Sole. The horizontal member of a frame wall or partition which rests on the floor.

Solenoid. An electrical device which closes when current passes through a magnetic coil.

Span. The distance between wall supports.

Specifications. A written set of instructions prepared by the architect covering materials, procedures, quality of workmanship and guarantees.

Split Ring. A metal device used at joints in wood trusses used to keep the members in position.

Stile. Side vertical parts of a door or window sash.

Stool. The shelf-like piece which crosses the bottom of a window on the inside.

Stop. The inside molding or piece of trim fastened to the jamb which holds the bottom sash in place on a double hung window.

Stucco. A cement exterior coating for buildings.

Studs. Vertical structural uprights which make up the walls and partitions in a frame building.

Sump Pump. A pump used to remove water from a sump pit sunk in the basement floor.

Sweated Joint. A manner of joining copper tubing using heat and solder.

Termites. Wood devouring white ants.

Terra Cotta. A ceramic material molded into masonry units before it is baked.

Terrazzo. A floor material consisting of broken marble chips bedded in concrete which is polished to a smooth surface after it is laid.

Thermostat. A heat sensitive device which is mounted in the living space and which controls the turning on and off of the heating plant.

Threshold. A piece of material over which a door swings.

Tie. *(Masonry veneer)* A metal strip usel to tie the masonry wall to the wood sheathing. *(Concrete formwork)* Devices used to tie the two sides of a form together.

Transit. A surveyor's instrument used by builders to establish points and elevations. The transit operates in both the horizontal and vertical planes.

Transition Piece. A sheet metal device shaped to form a transition from one shaped duct to another shaped duct.

Trap. A water seal in a sewage system to prevent gas from entering the building.

Tread. The horizontal board in a stair on which a person walks.

Truss. A structural member with primary and secondary members arranged to form a triangular assembly. Trusses are spaced at intervals to support the roof.

Valley. The internal angle formed by two inclined sides of a roof.

Veneer. *(Masonry)* A facing of stone, brick or other masonry material placed over a frame superstructure. *(Carpentry)* A thin layer of wood.

Vent Pipe. A small ventilating pipe extending from each fixture to the vent stack.

Vent Stack. A vertical pipe connecting with plumbing vent pipes and soil and waste stacks to carry off gases and to relieve pressure in the system.

Wale. Horizontal member which holds concrete forms in line and provides a stiffening effect.

Waste Stack. Vertical pipe to carry waste water to house drain.

Water Closet. A fixture usually called a toilet.

Water Table. The members of wood trim at the bottom of exterior siding designed for finish and to keep water from running down the foundation wall.

Weep Holes. Holes left between masonry units to allow water to escape.

Wythe. A single thickness of brick or stone in a masonry wall.

ANSWERS TO SELF-CHECK QUIZ

SELF-CHECK QUIZ NO. 1

1. Proceed with working drawings
2. Reproductions of architect's plans
3. Strength, translucency
4. Floor plans, elevation drawings, section drawings, plot plan, detail drawings
5. Rinse with water, fixing bath of potassium dichromate, a second rinse and drying
6. Van Dyke or brownprint
7. Surface of paper is moistened
8. Ammonia vapor
9. Blue, black or sepia
10. Easier to read, easier to make notations and corrections

SELF-CHECK QUIZ NO. 3C

1. L-shape
2. Gable roof
3. Asphalt shingles
4. Five inches rise per foot of run
5. (house) screened vent; (garage) screened louver
6. 2'-6"
7. Flush door with light
8. Panel door with divided lights
9. 16'-0" wide, 7'-0" high
10. Double hung
11. Each glass is 30 inches wide by 24 inches high and is divided
12. Double hung
13. Horizontal sliding
14. Insulating glass
15. Fixed sash
16. Hopper window
17. Bottom
18. 3'-2", 1'-10", 8'-0", 3'-0"
19. 3/8" exterior plywood with 3/4" x 1 1/2" battens
20. 8"
21. Face brick
22. Concrete
23. 16 x 8 vents
24. At chimney
25. Two

SELF-CHECK QUIZ NO. 3A

1. BRICK
2. BUILDING
3. CAST IRON
4. CEMENT
5. CEILING
6. COLUMN
7. CONCRETE
8. COPPER
9. CORNER
10. DIAMETER
11. DIMENSION
12. DITTO
13. DIVIDED
14. DOUBLE HUNG WINDOW
15. DOWN
16. DOWN SPOUT
17. DRAWING
18. ELEVATION
19. ENTRANCE
20. EXTERIOR
21. FINISHED
22. FLASHING
23. CONCRETE FLOOR
24. FEET
25. GALVANIZED IRON
26. GLASS
27. GRADE
28. HEIGHT
29. HIGH POINT
30. INCHES
31. LENGTH
32. LIGHT
33. BUILDING LINE
34. LONG
35. LOW POINT
36. METAL
37. MOLDING
38. MULLION
39. OUTSIDE DIAMETER
40. PANEL DOOR
41. PLATE GLASS
42. PLATE HEIGHT
43. RADIUS
44. ROOFING
45. ROUGH
46. SCALE
47. SHEATHING
48. SHEET
49. SHIPLAP
50. SPECIFICATION

SELF-CHECK QUIZ NO. 4

1. Not two full floors
2. Yes
3. 5
4. No
5. Yes
6. No
7. Dashed line
8. Yes
9. Yes
10. Yes
11. 2
12. Two dashed lines
13. No
14. 11'-4" x 23'-0"
15. Stone
16. 4
17. Furnace and fireplace
18. 4
19. Bedroom 3
20. No

SELF-CHECK QUIZ NO. 3B

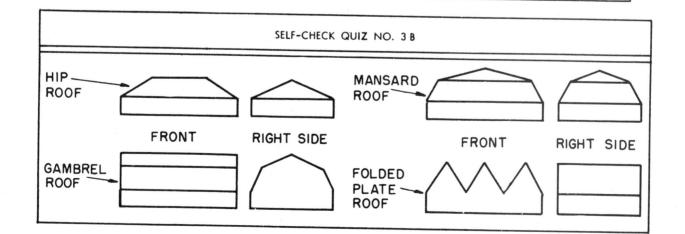

ANSWERS TO SELF-CHECK QUIZ

SELF-CHECK QUIZ NO. 5A

1. Common brick
2. Plaster line (lath and plaster)
3. Wood stud partition
4. Face brick
5. Wood stud wall
6. Brick veneer wall
7. Tile partition
8. Insulation
9. Cut stone
10. Fire brick
11. Tile
12. Cut stone on common brick (lath and plaster)
13. Stucco on frame
14. Solid plaster partition
15. Rubble stone
16. Common brick on concrete block wall
17. Glass block
18. Brick wall (no plaster)
19. Concrete
20. Concrete block

SELF-CHECK QUIZ NO. 5C

1. Living room, bath #2, bedroom #2
2. 5
3. Hall, den, kitchen, utility room
4. No. Garage down 3 risers
5. Brick veneer
6. Brick on concrete block
7. Wood stud partitions
8. Solid plaster
9. ∠ iron and ∠ iron and plate
10. WI pipe
11. East, west
12. Shorter
13. 16" OC
14. 4
15. Frame
16. Utility room and garage
17. Side lights to front door
18. Bath #2
19. Swinging, sliding, pocket, accordion
20. Hidden (dashed)

SELF-CHECK QUIZ NO. 5D

1. In utility room
2. Toilets
3. 3
4. Laundry tray
5. To get at piping
6. Over lavatories
7. Forced warm air
8. 9 WA, 5 CA
9. Baths #1 and 2, kitchen (utility room)
10. To counteract cold walls
11. Furnace provides heat
12. No
13. No
14. Yes
15. 6; change in flooring material from one room to another
16. 2 shelves and hamper
17. Windbreak
18. Rubber tile
19. Ceramic tile
20. Oak floor

SELF-CHECK QUIZ NO. 5B

1. ALUMINUM
2. CENTER TO CENTER
3. FINISHED FLOOR
4. INSULATION
5. RECESSED RADIATOR
6. WATER CLOSET
7. CATCH BASIN
8. COLD WATER
9. ELECTRIC PANEL
10. MASONRY OPENING
11. SHELF AND ROD
12. UNEXCAVATED
13. CEMENT FLOOR
14. CONCRETE BLOCK
15. EXCAVATE
16. PULL SWITCH
17. SINGLE STRENGTH GLASS
18. CENTER
19. EXPANSION JOINT
20. MEDICINE CABINET
21. SOIL PIPE
22. BEAM
23. CERAMIC
24. INTERIOR
25. OBSCURE GLASS
26. REFRIGERATOR
27. BEARING PLATE
28. FULL SIZE DETAIL
29. REGISTER
30. SWITCH
31. CINDER BLOCK
32. DOUBLE ACTING
33. RISER
34. CLEAN OUT DOOR
35. DOUBLE STRENGTH GLASS
36. GALVANIZED IRON
37. LAVATORY
38. PLASTER
39. RUBBER
40. BRONZE THRESHOLD
41. COLD AIR
42. HOSE BIBB
43. PLATE
44. WIDE FLANGE
45. DRAIN
46. HARDWARE
47. LINOLEUM
48. PLASTERED OPENING
49. TONGUE AND GROOVE
50. WEATHER STRIPPING

ANSWERS TO SELF-CHECK QUIZ

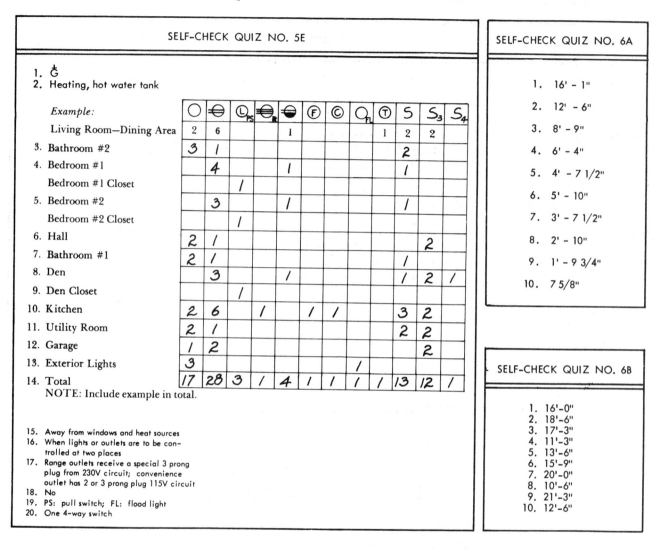

SELF-CHECK QUIZ NO. 5E

1. G̃
2. Heating, hot water tank

	◯	⊖	L_PS	⊖_R	⊖	F	C	◯_FL	T	S	S_3	S_4
Example: Living Room—Dining Area	2	6			1				1	2	2	
3. Bathroom #2	3	1								2		
4. Bedroom #1		4			1					1		
Bedroom #1 Closet			1									
5. Bedroom #2		3			1					1		
Bedroom #2 Closet			1									
6. Hall	2	1								2		
7. Bathroom #1	2	1								1		
8. Den		3			1					1	2	1
9. Den Closet			1									
10. Kitchen	2	6		1		1	1			3	2	
11. Utility Room	2	1								2	2	
12. Garage	1	2								2		
13. Exterior Lights	3							1				
14. Total	17	28	3	1	4	1	1	1	1	13	12	1

NOTE: Include example in total.

15. Away from windows and heat sources
16. When lights or outlets are to be controlled at two places
17. Range outlets receive a special 3 prong plug from 230V circuit; convenience outlet has 2 or 3 prong plug 115V circuit
18. No
19. PS: pull switch; FL: flood light
20. One 4-way switch

SELF-CHECK QUIZ NO. 6A

1. 16' - 1"
2. 12' - 6"
3. 8' - 9"
4. 6' - 4"
5. 4' - 7 1/2"
6. 5' - 10"
7. 3' - 7 1/2"
8. 2' - 10"
9. 1' - 9 3/4"
10. 7 5/8"

SELF-CHECK QUIZ NO. 6B

1. 16'-0"
2. 18'-6"
3. 17'-3"
4. 11'-3"
5. 13'-6"
6. 15'-9"
7. 20'-0"
8. 10'-6"
9. 21'-3"
10. 12'-6"

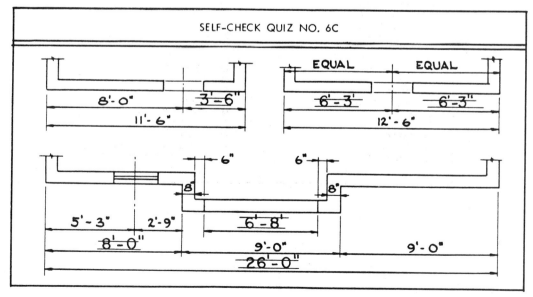

SELF-CHECK QUIZ NO. 6C

ANSWERS TO SELF-CHECK QUIZ

SELF-CHECK QUIZ NO. 7

1. 13
2. 11
3. 2
4. 12
5. 1
6. 10
7. 14
8. 3
9. 15
10. 9
11. 16
12. 4
13. 8
14. 17
15. 19
16. 5
17. 7
18. 18
19. 6
20. 20

SELF-CHECK QUIZ NO. 8

1. Gypsum board
2. Header
3. Inside casing
4. Stop
5. Jamb
6. Parting strip
7. Sash top rail
8. Sheathing
9. Siding
10. Drip cap
11. Outside casing
12. Meeting rail
13. Sash stile
14. Studs
15. Blind stop
16. Stool
17. Apron
18. Sill
19. Sash bottom rail
20. Glass light

SELF-CHECK QUIZ NO. 9

1. 90 x 160
2. 90 degrees
3. North-south
4. 40'-0" x 65'-0"
5. 6'-0"
6. 20'-0"
7. 100'-0"
8. 112, 108, 111.2 and 114.9
9. 6.9
10. SW
11. 114'-0"
12. 113.5
13. 1.4'
14. 1'-0"
15. Storm sewer, sanitary sewer, water main, power line

SELF-CHECK QUIZ NO. 10

1. National Electrical Code
2. Circuits
3. Ground
4. Flexibility
5. Threading
6. Waste
7. House drain
8. Trap
9. Equalize pressure
10. Bacterial action
11. Elevation
12. Riveting, soldering, welding
13. Flashing
14. Counterflashing
15. Gutter
16. Outside
17. Condenser
18. Valves
19. Floor, ceiling
20. Resistance
21. Point of beginning
22. Wales
23. Elevation, slope
24. Doors, windows
25. Bond
26. Welding
27. Shop
28. High pressure
29. Filler rod
30. Electrode, work

SELF-CHECK QUIZ IN MATHEMATICS

1. 2,293
2. 85,265
3. 312
4. 8,474
5. 10,192
6. 6,044,052
7. 21
8. 195
9. 1 7/8
10. 1 9/16
11. 20 3/16
12. 4/8 or 1/2
13. 2 7/16
14. 21/32
15. 2 43/64
16. 3 3/8
17. 84
18. 1/84
19. 12 1/4
20. 20.45
21. 317.132
22. 114.16
23. 1176.709
24. 1687.5
25. 19.125
26. 3.31571
27. 4.53
28. 9
29. 90.356+
30. .19